From Storeroom to Boardroom makes fascinating reading. It shows what is possible when you combine brains with integrity and courage. Truly inspiring!

Dr Okechukwu Enelamah,
Former Minister for Trade and Investment Nigeria (2015–19)

Babs' family gave him a strong ethical and educational footing, and more. This guided and helped him to overcome prejudices and negative perceptions to reach senior positions in some of the most challenging jobs around the world at Royal Dutch Shell. His integrity and business acumen were fully tested when he became CEO of a major gas company in Nigeria. This book is a must-read for business people working in challenging jobs and environments.

Ann Pickard, former senior executive of Mobil and Shell; non-executive director of Woodside, KBR, Chief Executive Women and previously Westpac and Catalyst; Fortune *magazine's Bravest Woman in Oil*

This is a story and unvarnished account of an authentic young man thriving amidst challenges… the author has conveyed in lucid prose his journey from the bottom to the top, aided by good mentors, coaches and sponsors who recognized his diligence, tact and integrity. The book is not a whitewash but an account of authenticity through the vicissitudes of life and how, through values of hard-work, integrity, humility and problem-solving skills, he rose from the storeroom to head Africa's best-run organization.

Clement Baiye, Commissioner, Nigeria Communications Commission (NCC)

Instead of complaining that a university entrance form did not allow enough characters to record his Nigerian name in full, Babatunde moulded the shortened version he was presented with – 'Baba' (meaning 'old man') – to his liking!

This anecdote about how Babs got his name illustrates the approach he enacted throughout his productive and impactful career: change what you can, take charge of your own destiny, compromise on the 'small stuff' but never on integrity and honesty.

Babs' memoir, rich in illuminating and amusing anecdotes and life lessons, is a refreshing story rebalancing our perceptions of Big Oil in Nigeria by showing how multinational-led joint ventures in oil and gas – and courageous personal leadership – help community development, act as a source of aspiration and growth for young Nigerians, and challenge and reshape a corrupt status quo.

Pamela Watson, author of Gibbous Moon Over Lagos: Pursuing a Dream on Africa's Wild Side

FROM STOREROOM TO BOARDROOM

How integrity and courage shape global business

BABS OMOTOWA

First published in Great Britain by Practical Inspiration Publishing, 2021

© Babs Omotowa, 2021

The moral rights of the author have been asserted

ISBN   9781788602341   (print)
            9781788602334   (epub)
            9781788602327   (mobi)

Practical Inspiration
Publishing

Contents

Foreword

I am delighted to be writing the foreword for this book on Babs Omotowa and leadership.

What is the relationship between them? Let me first ask the question: Are leaders born or are they made?

It is not one or the other, but a combination. Natural abilities are a great start, but it also requires nurturing through education, experience and coaching. The lesson learnt from childhood, from parents, role models, in faith, at school and from vocations like farming enables the inculcation of core values that become beacons of light, guiding leaders in later life.

The world has become a global village that has brought forth the need for the development of societies to bring people out of poverty and for a more equal world. While governments remain accountable to citizens for the development of societies, corporate leaders do have a role to play in helping address big societal issues.

Nigeria's journey to becoming a developed nation has been ongoing, beginning even before its independence in 1960. It includes partnerships between government and corporate businesses. One that holds a special place for me is the Nigeria Liquefied Natural Gas Limited (NLNG), conceived during my tenure as Head of State from 1966 to 1975. I visited the company in 2013 on Babs' invitation and I was impressed with the progress made by the company and its leadership towards the vision. I am delighted that the collective effort since then has further led to the final investment decision on the company's growth aspirations – the Train 7 project.

The authenticities of Babs' leadership at NLNG and later at Shell global headquarters in the Netherlands, are testimony to his core values of integrity and courage. This book provides excellent lessons for businesses and leaders, and serves as an incisive view of the varied challenges faced by the oil and gas industry, and how core values can help make a difference.

As we look towards an egalitarian world, and as businesses increasingly seek to be a force for good in the light of the lessons from the past, this book couldn't be more timely. It highlights how environmental, social and governance issues belong on the corporate dashboard, and how they can be addressed. It describes how by developing relationships with personal and corporate integrity can make a difference, enabling someone to thrive through challenges and create equality of opportunities.

I hope this fascinating and incredible story inspires many to do greater things in their endeavours towards making their country and the world around us a better place.

Kudos, Babs.

General Dr Yakubu Gowon, GCFR
Former Head of State of Nigeria (1966–75)
Chairperson Organisation of African Unity (1973–74)

Acknowledgements

I thank Almighty God for my life, experiences and opportunity to write this book. To Him be ALL glory!

I am grateful to my late parents, Chief Joseph Tolorunleke Omotowa and Dr (Mrs) Margaret Ebunolu Omotowa, for their legacies and legendary sacrifices, and for being my inspiration and lodestar.

To my brothers, Dele, Bola, Seyi, and sister, Atinuke, for their unending love over the years.

To my family, Helen, Mayowa, Titobiloluwa, Fiyinfoluwa and Oluwadara, for bringing me joy.

To my relatives and friends (Uncle Reuben, Biodun, Hakeem, Claude) for their endless kindness.

To my colleagues (bosses, peers, staff) at Shell in Warri, Aberdeen, Lagos, Port Harcourt and The Hague, who taught and supported me so much, for their great friendship.

To the staff at NLNG, where I had the opportunity to work with many talented and hardworking people, which gives me great hope for the future of Nigeria.

To my colleagues at CIPS and staff at Eason House, who helped me to build my professional knowledge, competence and global networks.

To the many friends who helped pre-read this manuscript – Ann, Pamela, Tony, Fola, Leye, Kudo, Pastor Mark and Arinze – and whose comments and suggestions made a world of difference.

Special thanks to Ifeanyi Mbanefo, who was with me from the beginning, for the invaluable guidance, encouragement and teaching on writing styles and advice on publishing.

A story is told of a pilot's fighter plane that was destroyed by a missile during combat. He ejected and parachuted safely.

Five years later, he was in a restaurant when a man came and said, 'You are the pilot who was shot down!'

'How in the world did you know that?' the pilot asked.

'I packed your parachute before you flew,' the man smiled in response.

The pilot gasped in surprise and gratitude and thought, 'If that parachute hadn't worked, I wouldn't be here.'

The pilot couldn't sleep that night, wondering how many times he might have seen the man and not even said, 'Good morning, how are you?' because he was a fighter pilot and that person was just a lowly safety worker.

My enormous gratitude goes to those who have packed my parachutes over the years and provided everything that made it possible for me to make it through the years. Thank you for letting me stand on your giant shoulders!

About the author

Cogito, ergo sum (I think, therefore I am)
– Rene Descartes (1596–1650)

It is a cold December morning in 2011. Babs gazes at a picture in a meeting room of Nigeria LNG (NLNG)[1] in Bonny Island, deep in Nigeria's Niger Delta. The photograph, taken 13 years earlier, was of NLNG's first liquid gas shipment from Bonny to Montoir in France. With that shipment, Nigeria had joined the league of gas-exporting nations.

He had flown from Lagos and driven into the Nigeria LNG Limited (NLNG) complex on the Island. This was the largest industrial plant in Sub-Saharan Africa, which produced 10% of the world's liquid natural gas (LNG) and had generated US$55 billion from its inception in 1999 until 2011. NLNG is owned by Nigeria National Petroleum Corporation (NNPC – 49%), Shell (25.6%), Total (15%) and Eni (10.4%).

In the meeting room earlier that morning, NLNG Board had appointed Babs as Managing Director and CEO of NLNG and as the Vice President of Bonny Gas Transport (BGT). With those appointments, he became a member of the board of directors of the two international companies.

As he looks through the window, Babs can make out a forklift in the distance, moving materials into a store. Babs takes a deep breath, recalling his first job, working for the Shell Petroleum Development Company (SPDC)[2] in a storeroom in Warri, Nigeria. He had never imagined that from those humble beginnings in the storeroom he would one day be in a boardroom as a director. It was all quite surreal.

---

[1] See www.nigerialng.com/Pages/index.aspx
[2] See www.shell.com.ng

Babatunde (Babs) Jolayemi Omotowa was born to farmer Joseph Tolorunleke Omotowa and educator Margaret Ebunolu Owonibi as the third of their five children. The first four children were boys, Dele, Bola, Babs and Seyi, who were born in the 1960s; the last child, and only daughter, Atinuke, was born in the 1980s.

Babs' father, Joseph, was a chip off the old block, as his own father, Pa Samuel, was a successful merchant of cocoa and coffee in his native village of Okoro-Gbede. Pa Samuel had seven wives and 25 children, with Joseph being the first child of his mother, who was the first wife. Pa Samuel's high number of children was normal in early twentieth-century Nigeria, as children provided free labour on the farms and reflected proactive planning against poor medical standards and the prevailing high child mortality rate.

Babs' mother, Margaret, was the first of three children of Pa Benjamin Owonibi and Madam Leah Owonibi from Aiyetoro-Gbede, which is situated 21 kilometres west of Okoro-Gbede.[3] She was a princess of the Owonibi dynasty: her uncle was the first regional king, Olujumu of Ijumu, HRH Oba Jacob Owonibi (1954–80).

The family's roots are in the Gbede communities of Kogi State, Nigeria. The Gbede communities are part of Ìjùmú land, inhabited by approximately 150,000 people who speak the Okun dialect. They migrated from Ile-Ife, the cradle of the Yoruba ethnic group in west of Nigeria and were predominantly farmers. Gbede is at the intersection of the forest regions and savannah plains of Middle Belt of Nigeria. Hills and rocky green habitats surround the villages, and blend to form a beautiful vista. The vegetation is a mix of the tropical rainforest of the Southwest region and the tropical grasslands of the Northern region.

---

[3] See www.maphill.com/nigeria/kogi/ijumu/aiyetoro-gbede/location-maps/physical-map

Map of Africa showing Nigeria

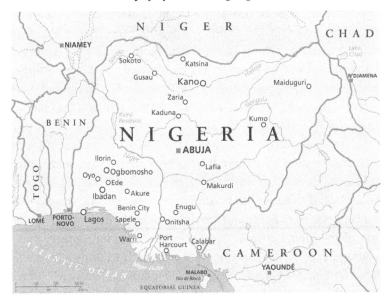

Map of Nigeria

Babs studied at the University of Ilorin in Nigeria and the University of Leicester in United Kingdom, where he obtained a degree in industrial chemistry and subsequently earned two Masters of Business Administration degrees (1994 and 2002), focused on operations research and supply chain management.

He began his professional career in 1989 as a science teacher at Bishop Smith Memorial College (BSMC) in Ilorin and later joined Nigeria Security Printing and Minting Company (NSPMC) in Lagos as a commercial officer. In 1993, he joined SPDC in Warri as a management trainee.

Babs began work in the storeroom of the SPDC warehouse in Warri and rose rapidly through the ranks in Shell ventures in Nigeria and Europe. He was, at various times, a Director of SPDC; Vice-President at Shell Sub-Sahara Africa; and a Non-Executive Director at West African Gas and Pipeline Company (WAGPO) and Nigeria Economic Summit Group (NESG). In 2014 and 2015, he was Global President for the UK Chartered Institute of Procurement and Supply, London.

After five years as MD/CEO of NLNG, in 2016 Babs was appointed to the Shell Global Upstream Leadership Team as a Vice President, based in the company's headquarters in The Hague in the Netherlands. The role, reporting to the Upstream Director, covered 40 countries across all continents and with responsibilities for climate change, greenhouse gas emissions, process and asset integrity, and the personal safety of 100,000 staff and contractors in the Shell Upstream business.

From his introverted days as a youth to teaching young students, to the endless wait for job opportunities, to being holed up in a storeroom, to working internationally, to tackling some of societal challenges and to becoming responsible for multi-billion-dollar decisions, Babs' journey from humble beginnings to the top of the global corporation took determination, integrity and courage. His life did not follow a straight line, and none of his achievements was predictable. Not only Babs, but many who knew him in his younger days, were surprised at what followed and who he eventually became.

Many people from developing countries may not believe that they can navigate through major corporations and follow the same trajectory into leadership as those from the developed world. Some may think this would be naïve, based on their own circumstances and experiences; however, the reality is that there are several such stories. This was one reason why, in the first interview after his appointment to the boardroom, Babs stated, 'There are no impossibilities.'

Introduction

Experience teaches only the teachable.
– Aldous Huxley (1894–1963)

It was midnight on 27 June 2019, and my penultimate night in The Hague. I had just returned from another dinner that had graciously been organized by colleagues to send me off. I was retiring early, after working for 26 years at Shell. I had decided to pursue other personal aspirations after a career in which I rose to the top echelons of one of the world's largest multinational corporations. My colleagues that evening had shared stories of many of our good times together. Recalling that I sometimes was the contrarian in the team and had many difficult encounters with co-workers, I wondered whether I deserved the compliments I received that night!

I had lived in The Hague for three years in a small flat on the eighth floor, atop a high-rise apartment on Carnegielaan Street, behind the Peace Palace, which houses the International Court of Justice. The apartment had great views and I often found myself at the window, soaking up the beautiful scenery of the city and reflecting on its rich history.

As I gazed through the window that night, my gaze darted across the brightly lit streets and rested on the canal (Mauritskade). The Netherlands is famous for its canals. With a quarter of its land below sea level, canals play a prominent role in the country's environmental security. They manage overflows of seawater and keep the cities from being flooded. The canals surrounding the cities used to serve as moats to protect them from foreign invaders. Today, they support tourism, irrigation, recreation and transportation.

I thought of Obalemo and Pere, streams in my hometown of Okoro-Gbede. In contrast to the Netherlands canals, Obalemo and Pere are unplanned and covered by bushes. They do not mingle with people and traffic, do not criss-cross the town. Yet they provide sustenance for the community – through fishing and irrigation, and as drains for stormwater.

Memories of my youth are full of excitement from running across dust-bowl paths, through the green vegetation and into the hills, to reach the slow-moving streams. As a child in the 1970s, my siblings and I frequently travelled with our parents to the village. Then there was no pipe-borne water in the village as underground wells and boreholes only reached the community in the 1980s. We often went to the stream, from which we fetched water in metal buckets, carried on our heads. It took several trips to fill the large drums in the house.

Eventually, I walked away from the window to my bedroom to get some sleep. My thoughts drifted to the plan I had been considering for many years – to write a book after retirement. During the various send-off dinners, colleagues had asked about my post-retirement plan. When I mentioned that I would write a book, many asked whether it would be an autobiography. I had always associated autobiographies with fame, wisdom, age, achieving records or sharing momentous events. I thought about it as I lay on my bed: none of these was me.

As I drifted off to sleep, I reflected on how magical it is when ordinary circumstances and everyday events hold hidden deeper truths and future implications. Many of the failures and experiences that occurred during my younger years later revealed their significance in my adult experiences and challenges. I intended this book to be about reflection on the key lessons that were critical in my journey.

Yet I would not be content with simply writing a book on my failures, challenges and successes. I wanted to write about episodes from my life, the lessons I had learned and the deep insights that I had gained during my career. I wanted to tell my unusual story – 'from storeroom to boardroom', from humble beginnings in a storeroom to becoming a CEO, and later a global Vice President of a major multinational corporation.

I would share the teachable moments of my life. The insights included the traits that had enabled me to reach great heights, the challenges I had encountered, how I responded and lessons I had learned. While my experiences are not universal, they provide insights into the characteristics that enable success.

My story shows that one can ascend in life irrespective of where one started from, and that opportunities are available when you have the right attitude. It demonstrates that education, hard work, creativity, courage, and developing character and integrity all pay dividends in the end.

My career was one of outright rejection of the status quo, and particularly the perception that Africans and Nigerians under-perform in global business. It was a career spent clambering up fences with bountiful rewards for determination, boldness and integrity. It required the courage to create a future of possibilities, ethical leadership in the face of adversity, focus and determination in the face of distractions, making an impact in difficult situations, building relationships and influencing others.

My story includes personal insights into the situation in Nigeria and the Niger Delta, and also into multinationals, which I hope will shed some light on some of the misconceptions that such companies are uncaring and unresponsive to the plights of communities and developing nations.

In managing big businesses in developing countries, one tends to see hot-button issues such as community development, corruption, pollution and local content only in the rear-view mirror, as secondary in the all-consuming chase for market and production targets. That is unless one makes a deliberate effort to put them on the dashboard or prioritize them on the corporate agenda. Readers will learn what it takes to run big businesses in developing countries and to deal with stakeholders.

I hope my story will be of interest to business executives and leaders, middle level managers, newcomers and students working in developing countries or planning to do so; to insiders and outsiders in the oil industry; to those with an interest in social issues; and to my compatriots. I hope that it will be relevant to businesses in developed nations with subsidiaries in developing countries or those considering establishing such a subsidiary, and to businesses in developing countries who encounter the same issues. It may prove useful, in some cases, to reinforce what many people already know, and in other cases, to provide new insights and reflection.

I hope that my story will inspire young people to be focused and not meander through life, but rather to show determination, courage and integrity as they face the tests of life. I hope the book will help in redirecting the steps of those who may have had core values instilled in them during their youth, but have lost their focus along the way.

I hope to share some of the monumental mistakes and personal failures that I made so that readers can avoid similar mistakes. I have had to dig deep to reach emotional levels that I have never explored before, resisting the temptation to omit stories that may not be flattering or consistent with who I am today.

By recalling moments of personal disappointment – such as my truancy in secondary school, which ultimately led me to repeat a year – I have been able to recognize my human imperfections. In the journey of life, there are no rehearsals and we all make mistakes.

I share some of my strategic failures, such as having to put a hold on a reorganization due to my failure to convince the board. Readers will learn about stakeholder sensitivities in complex environments.

I also share some challenges faced by the oil industry, such as climate change, community development, gas flaring and oil pollution of mangroves caused by pipeline corrosion, and mostly by sophisticated gangs stealing oil from pipelines. This will enable readers to gain further insights on the complexities experienced by companies operating in the Niger Delta.

I want to show how, with determination and courage, developing local capacity and communities can be achieved. I hope this will enable readers reflect on how to make a difference in the face of obstacles.

Further, I provide insights into the experience with an agency that blocked a company shipment to extort a levy, and in the process caused losses to both country and company. Readers will learn about the dilemmas faced by global multinationals working in developing countries that have different standards from their home nations.

The book also looks at some of the challenges of working in high-risk countries and how western views in European head offices may need to evolve in order to better tackle corruption in their overseas ventures.

I share some of the difficult relationships I have experienced with government officials and legislators who were insistent on amending or repealing legislation that was foundation of the company, and the efforts to preserve this.

I reveal how a growth opportunity with significantly benefits for the company and country was scuttled by political interests in government. Readers will learn how unrelenting efforts leads to success.

Importantly, I share some challenges of and insights generated by working outside my home country on international assignments in Europe, being part of a minority in an international company and how racial stereotypes played out.

I am extremely grateful to God for the opportunity to write this book with the objective of telling my story, in the hope that it will add to the knowledge of readers and inspire them to achieve greater success.

My story is one of gratitude to God for every step, from my parents, to my growing up and through my career. I owe it all to God, for His seen and unseen hand in my life. I am grateful to Him for the journey of my life.

Chapter 1

Beginnings

How can we know who we are and where we are going if we don't know anything about where we have come from and what we have been through?
– David McCullough (b. 1993)

Early missteps: Before success, experiences in my early years shaped me

It is dinner time at school, with all students (approximately 1000) eating their meals noisily in the big hall. The duty teacher, a boisterous man called Alhaji, shouts out my surname on the tannoy, and instructs that I should stand on the table. This is not a good sign, as it usually marks the beginning of a public humiliation and some punishment. I pay no attention to his instruction. The hall falls silent. I can hear the anger in his voice as he repeats, 'Omotowa, stand up on your table.' I continue to refuse. All eyes are on me. He walks up to me and repeats his instruction. I ask him, 'What have I done that I should be so humiliated?' He does not answer. The situation escalates as he pulls me by my shirt collar. I say to him, 'Alhaji, you are choking me', but he continues to tug. I become dizzy and feel that he is strangling me. At that moment, I decide that I need to free myself before I faint. I turn around, take a swing and hit him. The hall of students erupts in jeers. He is stunned, having been taken completely by surprise. He releases his hold on my collar. I adjust my shirt and walk out.

I knew there would be consequences but I felt I had been unjustly singled out and had not been informed of any offence. Alhaji reported the incident to the school authorities and without listening to any explanation from me, I was suspended from school. I deserved it. This was not good for me – especially having to explain what had happened to my parents. My father, not one to tolerate indiscipline, berated me, 'This is not how I raised you!' I felt sad at his admonition as I respected his opinion of me a lot. I was made to stay in my room and study throughout the suspension. I would only leave my room for meals.

I learned that one should not act on impulse but rather reflect on the consequences, to decide on options. This was one of the many lessons that I learned from experiences in my younger years – lessons that would come to be useful guides for me in later life.

This was Federal Government College (FGC) Ilorin, where I attended for my secondary education from 1978 after my primary education at Baboko Primary School. The government had conceived the FGC system in 1974 as part of its post-war effort to reintegrate various ethnic groups. The schools were designed as co-educational and unity institutions where students from all over Nigeria were selected on the basis of quota and merit, to ensure a broad student demographic.

Babs Omotowa (1978)

It was my first experience with such a diverse group, and it fostered a good environment to build lasting friendships and healthy networks, and to gain knowledge of other ethnic group, religions and cultures. I adjusted well to boarding school life. Coming from the home of strict disciplinarian parents, I was something of a model student in the early years. I was active in sports, particularly football (as a goalkeeper), and I was also an active member of the Christian Fellowship. I got good grades until the Third Year.

In the Fourth Year, I began to see the world beyond the strict boundaries my parents had built around me and my siblings. I became more adventurous and was anxious to explore. In a typical example, I became so fascinated with Brazil, for its football prowess and samba festivals, that I fantasized about stowing away on a ship. I shared this fantasy with my brothers, who still remember and occasionally joke about it. It was not until 25 years later, that I was able to visit Brazil. It was fascinating driving round Rio De Janeiro to see the Maracana stadium and eat local cuisine, *vatapa*, a delicious mixture of bread, shrimp, red pepper, ginger, peanuts and coconut milk. However, there was no samba festival happening, so that remains a fantasy.

Although an introvert, I was keen to spread my wings and try new things, which meant I did not resist the influence of friends at school who introduced me to truancy. On a particular day, I got formal permission to go to the dentist in town. A classmate, Ohis, asked me to buy him a packet of cigarettes. He told me where to buy them and how much they would cost. The shop attendant, unbothered by my school uniform, simply asked me which type I wanted to buy. I did not know the different brands and so asked the attendant for the most popular.

On giving the packet to Ohis, he reeled in laughter; the menthol cigarettes I had bought was viewed as one for 'softies'. He asked whether I wanted to try a cigarette. I followed him to a secluded part of school where I took my first drag. There he again laughed as I coughed and choked until I got the hang of it. I would go on to smoke for a few years before quitting.

From then on, I was invited to occasional smoking sessions and began to form closer bonds with the 'bad boys' in school. Later, I was invited to a lounge where I was introduced to alcohol. I gradually began to miss classes, and often left school without permission as a result of these habits.

My academic performance deteriorated. I went from being a student who was always at the top of the class to being mediocre. My grades continued to decline, but having been a high performer I took it for granted that I was going to do well at the end. I failed the final examinations and had to repeat the Fourth Year. I was shocked.

My parents were disappointed to say the least, considering their background and roots, and what education meant to them.

I learned that past successes were no guarantee of future success. I learned not to take things for granted or rest upon my laurels. I came to understand the impact of the people with whom one associates, and the influence they can have on one's performance, goals and ambitions.

My roots

My parents trained themselves through school, convinced that getting a good education was a lever to enable them break away from harsh realities of poverty in the village and from their ancestral agrarian background. They were dreamers who dreamt of the stars, of rising beyond their village environment and helping to uplift their community. They carried a torch of hope for a better future and understood that achieving their dreams would require them to travel to other developed societies to educate themselves and open them up to new ideas. They had tenacity and vivid purpose from early in life.

To realize his goal, my father worked manually as a farm hand to sponsor his education, including on his father's farms. He attended preschool and primary school in Kabba, a town next to his village. He then proceeded to Sapele College (170 kilometres away, in current Delta State) before attending Igbaja ECWA Teachers College (a mission school) in 1958. It was during this period at ECWA that he met my mother, who was still in her teenage years. She later attended Kabba Women Teachers College before they got married in 1963.

My father earned a scholarship to study at Forah Bay College (university) in Freetown, Sierra Leone (3000 kilometres away – a 47-hour road trip) and achieved a Bachelor's degree in history. He later received another scholarship, to the University of Durham in the United Kingdom, where he earned a Postgraduate Diploma in history. My mother also later earned a Bachelor's degree in fine arts from Ahmadu Bello University, Zaria, a Master's degree in art education from New York University (NYU) and a PhD in fine arts from the University of Ilorin.

My father (1980s)   *My mother's PhD graduation – Babs*
                              second from right

Education enabled them to achieve in their lives, building successful careers. In the 1960s, my father taught history and English at the Government Secondary School (GSS) Kano and GSS Keffi in the north. He became a principal at GSS Dekina and later at GSS Okene, then rose to become a Director of Education, a position from which he retired in 1985.

He was later appointed Chairman of Kwara State Education Board for four years (1986–90) to oversee hundreds of private and mission-owned primary and secondary schools and 25,000 teachers. He was reappointed to a second four-year term (1990–94), from which he was named Chairman of the newly created Kogi State Schools Board in 1991. Two years later, when he believed that he had achieved the objective of setting the board on a sound footing, he retired to his private and community development interests.

My mother taught at Teachers College, Ilorin, GSS Ilorin and Queen Elizabeth School, Ilorin. She became Vice Principal at GSS Tanke and later GSS Kabba, from which she retired in 1992. She was subsequently appointed a member of Kogi State Schools Board in Lokoja. She was hesitant to accept the appointment as she would be separated from her husband. Lokoja was 175 kilometres from Ilorin, where my father was living. My father encouraged her to accept the appointment, as he believed it was a recognition of her achievements, as well as an opportunity for her to develop and contribute to society.

My parents' appointments into board roles following their retirement were unsolicited. The recognition had come because they excelled at both their work and their personal development. I learned that focusing on excellence and personal development were key to advancement. Mostly, I learnt from my parents that education is worth more than any material inheritance parents could bequeath to their children.

Based on the role played by education in their lives, they espoused its value to everyone they came across. In the 1980s, my father had invested in a poultry business and experienced large losses. He turned to a certain Alhaji Abaye, a successful poultry farmer in Ilorin, for help. Abaye was a helpful consultant and through his support my father's poultry business became successful. They developed a close friendship.

After many visits to my father, Abaye – who was non-literate and had not previously considered the education of his children as important – told my father that he noticed that every young person in my father's home was studying. He interpreted the different circumstances as leading to different future directions for the two different families.

Abaye observed that although he was richer than my father, there was an air of confidence and sense of direction in my father's children that was lacking in his own children, which he ascribed to education. He asked my father how he got all his children to attend school and take studying seriously, as he wanted the same for his own family. He explained the predicaments of the cultural patterns of his native, social and religious background in contemporary Ilorin.

My father called on his first son, Dele, to guide Abaye's children. Eventually, two of Abaye's sons made it to university. One graduated with a Bachelor of Law degree from the University of Ilorin; the other graduated with a Bachelor of Arts degree in English from the University of Jos.

Despite Abaye being non-literate, he understood that education would lead to a better future and that his material wealth may not give his children such a future. His conviction that education creates hope for a better future reinforced my father's dreams.

Abaye's story is one from which anyone who may not have the lodestar that my parents provided could learn. Without a solid parenting foundation, it is important to identify in society those considered role models and reach out for mentorship. While there may not necessarily be a positive response from the first contact, with persistence it is possible to achieve success. Abaye changed the trajectory of his children's future by reaching out to others for mentorship. I was inspired by him, as it takes great humility and courage to seek help.

From my parents, I came to understand the role parents play in the journey of their children, through the values they espouse and their actions. Their examples remain useful guides for my relationship with my own children today. I also understood through my father the role that a husband must play in encouraging and supporting his wife to follow her dreams and reach great heights.

Parental values and senior examples are crucial in the development of children and the younger generation. As a result, it is vital that parents espouse worthy values and examples of good behaviour.

Retracing my steps

I would never forget my father's words when I presented him with my school report card where I had failed and been asked to repeat. He said calmly, 'Yemi, I am disappointed in you, you are better than this.'

I was called 'Yemi' throughout my primary and secondary school years. Most of my colleagues from that period still call me Yemi today. It was only from my university days that I began to be called 'Babs'. The genesis of Babs was during the completion of the form for the General School Certificate Examination. The column for one's name only provided for 21 characters. When I wrote my name, Omotowa Jolayemi Babatunde, there was insufficient room for all the letters. I inserted five more boxes to write all my names in full. However, when the official notification slip came back, it only provided for the 21 boxes, and listed my last name as 'Baba'. As Baba means 'Old man' in my dialect, I sought a different four-letter name and decided on 'Babs'.

After my school failure, it was clear that I had let the family down. My father was so upset that he did not speak to me for several days. I was withdrawn from the boarding house and commuted to school daily, and had no frills. I was not allowed any social interactions (e.g. TV, outings) and was restricted to my room. I tip-toed around the house and whenever my father returned from the office, I quickly disappeared to my room. The repercussions of failure provided me with a harsh and a bitter lesson.

I felt ashamed and this made me dissociate from my disruptive friends. I focused on my study and my grades improved. I finally graduated in 1984. While my grades were good, I did not make the grade to study medicine, my first choice. Instead, I was admitted to study industrial chemistry at the University of Ilorin.

I learned that to stand a chance, merely desiring a good outcome did not suffice. One needs hard work, perseverance and excellence in order to achieve and be successful in one's set goals. I learned to not be sloppy or distracted, or to spend time on irrelevant tasks, but rather to focus and work on what was important.

Beyond the lessons learned from my mis-steps in secondary school and then bouncing back, I hope this story offers parents whose children (despite being brought up in the best possible way) are in such a challenging stage as I was then with some hope that they can continue to trust God for the restoration of their children, and that the values they inculcated in the children at a younger age will help them through.

Introspecting ahead

At the University of Ilorin, my undergraduate years were filled with fun. I learned more about the country, people and cultures from excursions to mostly southwestern Nigerian towns such as Oshogbo, Ibadan, Iree and Ekiti. Prior to that time, I had not travelled beyond my home state. I improved my social skills through club activities and attending parties.

On one occasion, my friends and I learned of a party that was taking place in town. The students had talked about it for days on end. Despite our best efforts, however, we were not invited to the party. Nevertheless, we went at about midnight and tried to talk our way into the venue. We failed and were thrown out of the venue by the robust security that was in place. By 2.00 am it dawned on us that we would not get into the hall. Unfortunately, there were no taxis available at that time of the night and we could not walk to the campus, which was several kilometres away. We sat and dozed off on anything we could find. It was not until dawn that we got taxis back to the campus.

This was not a one-off occurrence. In another instance, I had gone out one day with a few friends. It was during the holidays and we spent the evening at a popular hangout. It was past midnight when I returned home. Our house was fenced and my father had the gates locked at 10.00 pm. After that time, the only way to get into the house unnoticed, and into my room in the small apartment at the back, was to climb over the 8-foot (2.4 metre) fence.

As I jumped down from the fence, I heard a distinct cough. It was my father, resting on a camp bed where he normally lay to gaze at the stars and reflect. This was also his way of getting fresh air, especially during electricity cuts. Blackouts from electricity cuts are a daily occurrence in Nigeria, as ~4000 MW of electricity capacity in the country is below 20% of what is required for residential and industrial needs.

That night, he was also waiting for me. He beckoned me and with head bowed I approached him in trepidation. I had been found out and I expected to be reprimanded. He inquired why I had been out this late at night. He asked whether I had reflected on the fact that there was a senior military officer living next door to our house and that any of his security guards could assume that I was an intruder and shoot at me. I had not contemplated such a scenario during the various other times in the past that I had jumped the wall.

I learned of the power of inquiry and self-reflection rather than articulation to make a more powerful impact. I had expected my father to let all hell loose on me, but his engaged approach achieved far more than fury and anger could have done. That my father asked me reflective questions rather than telling me off was unexpected, but it had the impact of leading to much better introspection.

I also learned about the need for holistic rather than narrow thinking about issues. I had been so focused on getting back into the house that I had not reflected on the other wider implications, such as being mistaken for an intruder and shot at. In the earlier story, we had not reflected on what would happen if we were not allowed in – especially to a party to which we were not invited. This learning was similar to the occasion when I took a swing at a teacher. It demonstrated to me that learning is not quite a neat, straight line. For every few steps that one takes forward, there is often a step back.

There were several such distractions during my university days. However, I learned to master my distractions from the FGC Ilorin days, and I did not allow the social distractions to negatively affect my education. I was not distracted by less important activities – a lesson that would stay with me throughout my career. I ensured that I read and studied adequately, and I graduated in 1988.

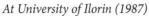

At University of Ilorin (1987)	*Family (1990), with two female relatives (Babs in suit)*

Standing up and implications

I could hear the commotion outside from inside the house roof where I was hiding. The distinct voice of the ringleader of the marauding students, bellowed, 'Bring him out or we will burn your house!' The landlord appealed to them. He tried to calm them. He explained that I had only run through his compound and that I was not hiding there. The rancorous voices gradually started to fade. I asked myself quietly, 'How did I get here?'

I was posted to Ondo State after graduation, for the compulsory one-year National Youth Service Corp (NYSC) and was assigned to a School in Ile-Oluji. NYSC was set up by the military government in 1973 as a compulsory programme to involve university and polytechnic graduates below 30 years of age in nation-building. It starts with one-month paramilitary training, supervised by soldiers and followed by an 11-month assignment in a government ministry, agency or private company.

During National Service (1988)

I enjoyed teaching at the school and made friends with the students and teachers. I also made several friends in the town through social interactions. I was fondly called 'lanky corper', which highlighted my tall and slim frame. 'Corper' is a nickname for members of the NYSC.

I was seven months into my service year when I was attacked and almost lost my life. I was teaching Fifth Year students one day when I heard loud shouts from the class next door. It turned out that some male students were harassing a female teacher.

On discovering this, I rushed there and stepped in between her and the students. I asked them to stop. My six feet and four inches height must have been intimidating and they dispersed. After rescuing the teacher, I escorted her to the staff room. At the staff meeting that followed, the unanimous recommendation was to expel the students. The school principal endorsed this and implemented it.

A few days later, I was the duty teacher, supervising some students cleaning the school compound. Suddenly, the expelled boys arrived at the school, with machetes and clubs. As they ran towards me, my assumption was that they wanted to scare me as it was broad day light and students around as witnesses. I asked, 'What are you trying to do?' I stood, expecting a verbal confrontation and threat, but I was wrong.

They attacked me and another teacher. I was hit by a club and I fell. As I looked up, one of them swung his cutlass aloft. I dodged but I was not successful, and I received a deep cut around the left eye.

I scrambled up and ran. Surviving today is always important in order to thrive tomorrow. I managed to escape with blood dripping from the incision. I jumped fences to reach a neighbouring homestead, where a family (named Olorunmodimu) hid me in their roof. The boys, suspected to be under the influence of alcohol and drugs (marijuana) due to their bloodshot eyes, embarked on a house-to-house search for me.

It was a terrifying wait in the roof, where I was also left wondering what fate may have befallen my colleague in the melee. The family in whose roof I had hidden later contacted the school principal, who eventually arrived with police to take me to the hospital for medical treatment, where I also met my colleague. The police later found and arrested the boys.

My friends telephoned my parents and two days later my mother picked me up, together with my belongings. The Ondo State Director of NYSC visited me at Ilorin with other officials. They came to appeal to my father not to press charges against the school as well as the commission responsible for posting me to the school. My father told them that as I was alive, he had no charges to press. I received a month of medical treatment under my parents' care.

My injuries healed during this period, although I was left with a scar across my eyebrow. I returned to attend the passing out ceremonies in Akure (Ondo state capital) in 1989. I was sad not to have had the opportunity to return to Ile-Oluji to say goodbye to the friends I had made while there. I continue to hold fond memories of the town. The attack was frightening, and I am most grateful to God for His protection.

I learned that courage and standing up for what is right (in this case, rescuing the female teacher) were important, despite the personal risk it entailed. Such is the sacrifice that may be required for any good cause. I also realized that given such risks, it is wise to ensure adequate protection and minimize unnecessary exposure.

I learned from my experience with marauding students that travelling such an unwavering path is a great attitude to take in life, but one that comes with a price. One cannot appreciate the good without enduring the bad, as without darkness one may not appreciate light. Difficult roads often lead to beautiful destinations.

Enabling others to grow

After my national service in 1989, I was employed to teach mathematics and chemistry at the Bishop Smith Memorial College (BSMC) at Ilorin, Kwara State. This was a continuation of the teaching experience I had during my service year, which I had thoroughly enjoyed. I put myself wholeheartedly into teaching, taking time to study widely to be able to impact my students with knowledge. I found that I needed to remain on top of my game to be able to teach properly, especially as the students were inquisitive and asked lots of questions.

The school principal took notice of my hard work and appointed me coordinator of the Examination Committee. Back then, the organization of school-wide exams at BSMC was a Herculean task, as there was no internet and no personal computers in use at that time. The coordinator was required to liaise with external partners to prepare the school and the students for the examinations. The committee that I led had to process, print and distribute examination papers.

The principal also appointed me as the school's Assistant Sports Master, and later the substantive Sports Master. In these additional roles I learned how to multi-task, which helped me to develop various abilities in working with people.

Later in life, some of my students served in government, business and the military. Once, on a visit to Nigeria's Presidential Villa (Aso Rock) to resolve some issues, a former student introduced himself and subsequently introduced me to other senior government officials. Those introductions enabled me gain helpful insights into the issues that had brought me to the president.

I consider teaching to be the most fulfilling job that I have ever done because it led to the transformation of people. The mentoring of young kids through their journey – from when students start in school, knowing little or nothing about some of the subjects, through to when they complete their studies, armed with knowledge and skills to go on and achieve greater things – is incredibly inspiring. I learned that nurturing others to succeed by sharing knowledge and experience can bring joy.

My parents as lodestars

I recall with nostalgia those humble beginnings and period of innocence. The early year experiences and lessons played a vital role in my later life as they provided me with great examples of values and characters that were important for success.

I learned from my roots. I was particularly in awe of how my parents, from such humble beginnings, had come so far. They followed their dreams and attained improbable heights. Watching and appreciating how they went about their achievements, I gained insights that eventually played key roles in my life and career.

The sheer determination to educate themselves, the hard work they put into farming to raise funds required and the courage they had to travel to far distance places in pursuit of better education taught me a lot about the value of education. It was what enabled them to escape their agrarian background and reach great career heights to build a better future. My mother's attainment, in the face of her other challenging family responsibilities, was an inspiration all through my school years.

My parents went through life with values of honesty, integrity and contentment. The virtues of contentment have stayed with me throughout life – for example, the last car that I personally owned, even as a Director at Shell Nigeria, was a small Kia Picanto. Contentment makes it easy to be honest.

I also learnt from observing my father's adherence to the core values that every strength has its weakness, as his insistence on integrity could sometimes be viewed as inflexible. As with knowing what is right, there are no shades of grey with integrity.

My parents were lodestars that I used to navigate my journey in life. I owe much of my start in life and in my career to their efforts – not just for the financial stability they provided, but for instilling the important values, character and attitude that they passed on to me. They are the giants on whose shoulders I stand, and their achievements helped me to see my own career successes as relatively modest. I learned from them the value of having a clear vision, determination, hard work and excellence. These were lessons I also learned during my school years, my missteps and bouncing back.

Lodestar parents are natural role models for children; conversely, it could be more challenging for those without such role model parents. In addition to identifying role models to reach out for mentorship, as in the earlier example of Abaye, reading books and listening to speeches can also provide inspiration on the characteristics and traits that will make a difference. Observing and learning from others provides endless opportunity.

Chapter 2

Determination

The difference between a successful person and others is not a lack of strength, not a lack of knowledge, but rather in a lack of will.

– Vince Lombardi Jr (1913–70)

Pursuing opportunities

It is a cold Harmattan morning[1] in Lagos in early January 1993. I show the employment letter I have just received by post from one of the largest multinational companies in Nigeria to my friend, Ade. Tears stream down his cheeks as he goes through the letter from SPDC. They are not tears of joy but rather of disappointment in himself. 'How did I miss out on this?' he asks aloud. I try to encourage him, saying there will be other opportunities in the future. He retorts, 'But this was an opportunity that was also available to me and I did not take it.' I say, 'Don't be too hard on yourself, but learn from this experience.' He replies, 'I should have applied when you did, and when you encouraged me to do the same two years ago. I should not have been so dismissive.' He finally composes himself and says, 'Congratulations, this is worth celebrating. Let us go out!'

Determination requires sheer grit to achieve a purpose. It is making up your mind about something and pursuing it to its

---

[1] The Harmattan is a season in West Africa that is characterized by the dusty Harmattan wind that blows in from the Sahara. It occurs between November and March each year.

logical conclusion without letting anything stop you. It is going for something against the grain and even against the expectations of others. It is firmness and not giving up despite obstacles and difficulties. It can be stressful, as you may meet obstacles on the way and need to do things you have never done. It needs absolute commitment and a laser-sharp focus on a set goal.

Having learned about the importance of determination when it comes to achieving success from both my parents' self-education and their careers, and my own schooling experience, I put the lesson to good use. It played a crucial role in enabling me to secure this job, and in persuading my bosses to develop confidence in me throughout my career.

Focusing on a goal

In the early 1990s, I had found teaching both exciting and fulfilling but also challenging. I was earning the equivalent of $100 per month. Had it not been for the fact that I was living in my father's apartment, rent free and being fed at home, I would not have been able to make ends meet – especially as I had also enrolled for a Master of Business Administration (MBA) degree at the University of Ilorin.

I could not wait to complete the Masters programme and seek greener pastures elsewhere. It was on a part-time basis, as I was also teaching at the secondary school at the same time. I completed my MBA and immediately resigned from my teaching job in 1991. Handing over my resignation letter was emotional as I had been a key member of staff. The principal had asked, 'What can we do to make you stay?' There was nothing, as my mind was made up. I was determined to pursue my goal of finding a higher-paying job in an organization where I could build a good career. I left Ilorin for Nigeria's then-capital, Lagos, in search of better employment.

In Lagos, I visited many places that posted job advertisements at the front of their buildings. I frantically searched for vacancies in daily and weekly newspapers. Friends and family sent me information on other job advertisements of which they became aware. I was convinced I could secure a job through these means, purely on merit. It was not uncommon during that period for jobs to be secured through connections, but my parents were only teachers and did not have such networks. They also did not believe it was right. They trusted more in the will of God.

I applied for hundreds of vacancies and attended many interviews. I recall once interviewing at a start-up commercial bank in Ibadan located 150 kilometres from Lagos. The bank offered me a job, but I walked away. The salary was not good enough in my assessment. While I was leaving the bank's premises, one of the company's officials ran after me. He encouraged me to accept the offer, and said that once the bank stabilized, it would be in a better position to increase my remuneration. One might have thought that, being unemployed at that time and with an offer of a salary twice what I was earning as a teacher, I should have accepted the job with prospects. However, my objective for leaving the teaching job at Ilorin was clear: to get significantly better-paid employment. I would not settle for anything less. I had considered the cost of accommodation, living expenses, transportation and what would be required to have a decent life and I was convinced that the salary offered was just not enough.

Living in Lagos, with its population of eight million people, was a new experience for me, having grown up in a small town. It has one of the busiest seaports in the continent and was Africa's fastest growing commercial city in the 1990s. Lagos is divided by a lagoon into two areas: the densely populated 'Mainland' and the affluent 'Island'.

The Island is the commercial centre, with many financial institutions in high-rise buildings. It has the major Idumota and Balogun markets, military barracks and federal establishments. The Island has the tallest skyline in Nigeria.

The Mainland, which is more residential, was where I lived with my friend Biodun at Surulere. We had lived in the same neighbourhood at Ilorin and we were school mates. His father was a biology professor at the University of Ilorin, and we lived in the annex of his father's bungalow on the university's main campus during our university days.

Weeks after arriving in Lagos, I visited my Uncle Reuben at his office at the Federal Secretariat in Ikoyi, on the island. I wanted to tell him I was in town so he could let me know if he knew of any vacancies. He was surprised to see me, unaware that I had left Ilorin. In those days there were no mobile phones and only a few landlines in Nigeria.

When I told him I was staying with my friend in Surulere, he was upset. He insisted that, as his nephew, I could not come to the city where he lived and stay with someone else. He asked that I arrange to move my things immediately to his residence in Onipanu.

Being a director in a federal establishment, I had assumed that my uncle would be living in a large apartment with several spare rooms. When I arrived, I was surprised to discover that it was a one-bedroom studio. He said that the one bedroom was large enough for both of us to sleep in. I was heartened – not only was he comfortable in showing me his modest apartment but had invited me to live with him. He did not use the apartment's size as an excuse. I knew this would be both an inconvenience and a sacrifice for him.

I lived in the apartment for many months, during which he made me comfortable. He ensured I had money to buy food as there was no kitchen in the house – I am not sure either of us could cook anyway. In the evenings, he gave me money to hang out or bought me drinks and pepper-soup (a Nigerian delicacy) in the local bar.

I was fortunate to meet friends and relatives who helped me along the way. Without their support, it would not have been easy to cope with the challenges. I remain grateful to them.

I learned that it is not how much one has that matters, but the willingness to share whatever little one does possess, as we must never under-estimate the difference it can make to another person's life.

Perseverance pays off

By December 1991, six months into my sojourn in Lagos, I had not yet received any job offers despite many applications. As it was approaching Christmas, I decided to travel to Ilorin to take a break and be with my parents and siblings during the holiday. I travelled in the company of my friend Ade, who had been my MBA classmate at the university and was also living in Lagos.

At the public motor park, we waited for the commercial bus to fill up with passengers. During the wait, I bought a copy of *The Guardian* and read the job vacancies section. There was an advertisement from SPDC, a joint venture of Royal Dutch Shell, which wanted to recruit graduates of various disciplines as management trainees.

I showed the advertisement to Ade, suggesting that we apply when we got to Ilorin. He commented that such companies reserved jobs for children of high and mighty influential people, not children of civil servants and teachers like us. I told him that we had nothing to lose and besides, this was an international company. Who knows, maybe SPDC's hiring process would recruit on the basis of merit? He laughed this off as we continued our journey.

By the next day, I had written my application. Thinking that we could post our applications together, I asked Ade if his was done. He laughed and said he did not have time to waste on such things. I tried to convince him that he still had an opportunity to apply. He was adamant and remained uninterested. I was disappointed.

We returned to Lagos after the New Year holiday and I continued my job search. A few months into the year, I received a response letter from SPDC, inviting me to attend a written test at their office in Port Harcourt, Rivers State.

Ade laughed when I showed him the letter and said, 'They are just using the likes of you to make up the numbers and give credibility to the process.' He insisted that those on the assessment panel had already made up their minds about who to employ and I was just wasting my time. I was determined to attend the written test and give it my best shot. I had not been to Port Harcourt or that part of the country before, so getting to know another part of Nigeria would also be a positive outcome of the exercise.

Uncle Reuben paid for my trip. He also gave me his friend's address at Central Bank (CBN) in Port Harcourt, with the instruction that I could stay with him for the duration of the test. I travelled on a night bus and the next day I took the test.

At the end of the test, the recruiters asked me to submit my expenses receipts for air travel and hotel accommodation. I had only come by road and did not stay in a hotel. They calculated the in lieu amounts for these and paid me what I considered a substantial reimbursement. On returning to Lagos, I let Ade know about the experience and how I was unexpectedly reimbursed beyond my expenses. This meant my trip was a financial gain. He shrugged and wished me well but restated his view that it was a wasted journey.

A few months later, I received another letter from SPDC asking me to return to Port Harcourt for another round of tests. I travelled for the test and this time I stayed in a hotel. I did not travel by air because commercial flight operations in Nigeria were not reliable then and I could not risk a cancelled flight.

By the summer of 1992 I was offered a job as a commercial officer with Nigeria Security Printing and Minting Company (NSPMC) in Victoria Island, Lagos. The company prints the nation's currency (Naira). It was one of the many hundreds of companies to which I had applied in the past. Unlike the previous employment offer that I had received from the bank in Ibadan, the salary offered by NSPMC was much more than expected and, as it was a reputable organization, I was glad to secure such a job.

Two months into the role, I was posted from Lagos to the Ibadan Office, covering the Western Region of Nigeria. This involved traveling to cities such as Benin, Ore and Ondo to follow up with the customers on printing of government security documents. This role exposed me to several clients and organizations, and I learned about relationship management.

A few months into my work at NSPMC, I received another letter from SPDC to attend an interview in Port Harcourt. Ade remained very sceptical about any positive outcome.

Uncle Reuben, who had been supportive so far, also felt that I could be wasting my time responding to the invitation for interview. He felt it was unnecessary to subject myself to the stress required to travel to an interview for a job that may not be any better than the one I had at NSPMC. He was also concerned that I may be putting my NSPMC job at risk by taking time off work so early in my employment. I had indeed not yet been confirmed at NSPMC at that time. I explained that as I had started the process with SPDC, and as I was making progress, I wanted to complete it and see the outcome. He was not convinced. I enlisted the support of other family members to convince him. In the end, he reluctantly agreed to support me.

I travelled to Port Harcourt for the third time to attend the interview. On this occasion, it was an oral interview and the panel was composed mainly of expatriate managers. Coincidentally, my MBA thesis had been on inventory management and as the questions were mainly on materials management, I was quite at ease with their line of questioning. The interview became more of an interactive session as we discussed many of the concepts involved in running effective inventory management and I shared some of my thesis findings.

In December 1992, I received a letter from SPDC, asking me to come to its Lagos office for medical tests. For the first time since the start of the process in December 1991, Ade was somewhat surprised. He told me that the medical test may be the point at which I would be dropped from further consideration. However, I attended, all went well and I returned to my work at Ibadan.

Early in January 1993, the employment offer from SPDC – the letter I had showed Ade – arrived by mail. I also showed my Uncle, who observed the significant difference in compensation compared with NSPMC. He was surprised and admitted naïvety about the emolument and potential career that the job at SPDC would afford me. He was glad that I had attended the interview despite his protestations, and he felt immensely proud. I was and remain very thankful to him, as he had played a major role in the process.

I resigned from NSPMC immediately, paying back the three months of salary stipulated in my contract. I was ecstatic and could not continue working at NSPMC for the three-month notice as I was eager to begin work at SPDC, an international conglomerate I had admired from a distance.

My determination paid off as I got my dream job meritoriously. To work for SPDC was a dream come true. This was achieved despite the distractions along the way. Resigning from the teaching job without securing other employment was moving from a known to an unknown. Rejecting the offer from the commercial bank was unusual. But these were risks I was willing to take in pursuit of my goal.

I learned that one must not be distracted by the cynicism of others. It is important to be clear and focused on a goal, approach it with determination and not settle for half measures.

This reinforced the lessons I had learned on my father's farm about the value of perseverance and determination.

Shell in Nigeria

Sir Marcus Samuel founded Shell Transport in 1897, operating tankers to the Far East. In 1890, the Royal Dutch Company for Exploitation of Oil Wells in Dutch Indies began tapping oil in Sumatra.[2] In 1903, Royal Dutch and Shell integrated operations and in 1907 merged to become the Royal Dutch Shell Group.

In 1936, the company started business in Nigeria as Shell D'Arcy.[3] It discovered oil in 1956, at Oloibiri in the south-east Niger Delta.[4] Shell entered a joint venture with the Nigeria Oil Company, Total and Eni, called Shell Petroleum Development Company (SPDC) and built 6000 kilometres of pipeline, 1000 producing wells and hundreds of flow-stations and gas plants.

When I joined SPDC, it was producing a million barrels of oil and gas each year, and had good relationships across the 180 communities in which it operated. The main challenges were gas

---

[2] See https://en.wikipedia.org/wiki/Royal_Dutch_Shell
[3] See https://en.wikipedia.org/wiki/Shell_Nigeria
[4] See www.arabianjbmr.com/pdfs/OM_VOL_2_(11)/4.pdf

flaring and under-development of the communities, a complex quagmire that led to unrest in the Niger Delta (see Chapter 5).

First insights into an oil city in Nigeria

The letter from SPDC had requested that I assume duties at the company's office in Warri, Delta State. I was full of excitement going to work for a joint venture of a well-known multinational. I took a taxi from Lagos for the six-hour journey. This was my first trip to the town bordering Sapele, where my father had attended school 40 years previously. The drive through Ore to Benin was smooth, but as we made our way on the last leg to Warri, the roads became worse and the journey was tiring. Most of that stretch of road was untarred with huge potholes, which made the ride uncomfortable. I was surprised and disappointed as I had expected the road into one of the main oil-producing cities to have been a multi-carriage highway.

Warri is a port town in the western part of the Niger Delta. It has a seaport for cargo between the Niger River and destinations across the Atlantic Ocean. Its population in 1993 was 200,000, mainly of the Urhobo, Itsekiri, Ijaws and Isokos ethnic groups, and with a large influx of people – especially the Ibos. Its history dates to the fifteenth century, when it was visited by Portuguese missionaries. By the eighteenth century, it was serving as a base for Portuguese and Dutch slave traders and it later became a centre for trade in palm oil, rubber and cocoa. Warri gained prominence during the 1960s civil war for its strategic importance to soldiers on both sides of the conflict due to its seaport and refinery. It housed oil companies such as Shell, Chevron and Agip, as well as a petrochemical plant and a steel plant. A large number of expatriates lived there in the 1980s and 1990s.

I was shocked at the visible under-development. Looking out of the taxi window, I was disturbed by the array of corroded zinc roofs, worn-out buildings and rickety wooden kiosks lining the streets. This was not the image of Warri that I had in my head. I had heard so much about the place, which was the second largest oil town after Port Harcourt. I wondered about the actual impact of the oil revenues generated over the years, from the crude oil and gas, the refinery and the seaport.

I reflected on what may have been the cause of the level of under-development of the town despite the huge sums of money that had been allocated by the government. The poor infrastructures and visibly abandoned projects provided a hint. The money must have been frittered away. The corruption that had stalled the development across the country was also prevailing in the town.

I learned at first hand of the paradox of the 'resource curse' and saw how the abundance of oil resources had failed to translate into the development of an oil-producing city like Warri. I was disappointed to see the city's inhabitants living in poor conditions, an indication of the failure of past governments.

We stopped at SPDC's guest house, located off Airport Road, where we were accommodated for three months. In the evening, I went for dinner and got to know other new employees, some of whom I had met during the medical test a few months earlier in Lagos.

Early the next morning, on our way to the office, the bus stopped abruptly on the highway. A barrier had been lowered across the road. A similar barrier was lowered 100 metres ahead to stop oncoming vehicles. I was still wondering what was amiss when I heard the noise of an aircraft. A small Twin-Otter propeller plane landed adjacent to the road in a small adjoining airfield. Most of us had never witnessed a plane landing over a highway. These planes provide transit between SPDC's offices in Lagos and Port Harcourt, and for connections to international flights.

At the office, we commenced a month-long induction in the materials department. The department was responsible for the procurement, storage and transportation of oil field materials and equipment (pipes, valves, flanges, cables, pumps, electrical, instruments, spares, fuel) for the company's oil operations in the western part of Nigeria. I was impressed with the onboarding process and the professionalism of the staff. I was assigned a personal computer – it was my first opportunity ever to use one.

Along with the other newly recruited colleagues, I had been under the impression that we were going to be posted to offices, in white-collar roles. We therefore dressed in suits and ties during our induction. We became the butt of jokes by some of the older staff in the department, who knew that our formal dress code would soon change to more informal attire. They were right: within a short period, I became extremely comfortable coming to work in coveralls. I learned to listen more to experienced people.

Starting at the bottom (the storeroom)

Abel walks into my office on a Monday morning in 1993. He is one of the staff working in the storeroom. He looks dejected and says to me, 'There has been a break-in and theft in the shed area.' I sigh and ask him, 'What is missing this time?' He replied, 'It is the stainless-steel valves we received into stock last week. I am sad.'

This follows a pattern of break-ins at the weekends when staff are away. I wince as I think of having to report on theft again at the supervisors' meeting later in the day. This is not what I expect to be grappling with in this role. 'Abel, let us go and have a look,' I say. I do not want to look demoralized in the presence of my team members, but these repeated thefts feel like a heavy weight on my shoulders. I am just getting familiar with aspects of this job, but it has been difficult to make the changes required. 'This cannot continue and we must bring it to a stop,' I say to the staff in my team who have gathered around us in the shed. They look at me with doubt written on their faces.

This is a low moment for me and the team, but it invigorates me with determination to drive the needed improvements, and to successfully start my journey from the storeroom.

This was my first assignment in SPDC as a supervisor of the storeroom (warehouse) where materials and equipment were stored. These were to be retrieved and used by field staff to operate, maintain, repair and replace production equipment in oil and gas fields. I was responsible for the receipt, storage and issuance of items; maintaining inventory; keeping record and reconciliations;

and ensuring the effectiveness of the warehousing activities and staff performance (20 direct and indirect staff).

I spent the first few months rolling up my sleeves and making efforts to understand the activities, processes and culture. I took time to assess the physical structures and layout, quality of staff, store record systems in use, processes for receipt and issuance of materials, and the culture and norms. I contextualized these against how a good storeroom operates. I studied warehousing, read reports such as audits and visited other company warehouses to learn. I spoke with staff, other supervisors and customers to understand the issues, challenges and opportunities.

I wanted to do this early on in my time at SPDC, having learned from my father's farming activities that planning, preparation and hard work are key to success. He continued farming during his career and spent most of his private time developing farmsteads on leased lands where he planted several crops.

Every Saturday while his four sons were on holidays, we accompanied him to the farm, to manually plant, weed and harvest crops, from dawn to midday, for over a decade. This enabled us to bond in many ways: walking to the farm, sharing food, the pain of completing assigned jobs, and spending idle moments in discussion.

My father was happy for us to be engrossed in the farm work, as that meant more accomplishment. It was also a way to teach us about planning, hard work, determination and reward, and to minimize idle time spent socializing.

Returning from the farm (Babs with basket in hand)

I understood the sequences and phases of farming, including clearing, ridging, planting, nurturing and harvesting. We first had to cut and clear the allotted land of any obstructions, including weeds and trees. We would then till the ground and make ridges. This enabled us to plant seedlings and to tend them until they sprouted.

We removed weeds to prevent overgrowth that would compete for soil nutrients and choke the crops. This continued until the crops reached maturity. The final phase was harvesting, after which we would take the crops back home and store them before the family either consumed or sold them. The various phases of farming were timed to coincide with the seasons. For example, planting needed to be completed before the rains started.

I learned about the need to plan, as achieving results required several steps to be planned and timed appropriately. Through the phases of nurturing, I learned that one had to invest time, demonstrate perseverance, and thereafter trust in the outcome of so much effort.

The storeroom, which housed hundreds of millions of dollars worth of items, was poorly organized. Materials and equipment were strewn all over the place. There was no logic to their location and arrangement, so it was a herculean task to find items when customers requested them. The stores were an open space with roof and bare floor. There were no walls. The chaotic situation was an opportunity for theft and items (especially the small instrument and electrical spares) were often pilfered. Intruders could easily access the materials in the open space, yet I suspected some of the staff might also be involved.

I was disappointed that the company had invested significant amounts in procuring materials and equipment for operating and maintaining its oil and gas facilities, yet failed to secure them over the years. The value of the items, as well as their use, justified such protection. They were mainly used for exploration and production of hydrocarbons, and their operation and maintenance were the core activities of SPDC. We always reported the theft to the police and, through their investigations, we discovered that the items that were stolen were sold on the black market, mainly to contractors, who in turn sold them to other oil companies or to manufacturing industries.

I discussed my observations, and the challenges, with the staff working in the storeroom over several weeks. All were concerned with the negative image that the storeroom had in the company. It was viewed as a 'junkyard'. This was not something of which we were proud, so we all agreed that there was an urgent need for change. The staff had over the years believed there was need for such turnaround, but they had assumed that the company management was not interested, since they were hardly ever visited by senior leaders.

All these reviews and engagements enabled me to identify what I considered the top three areas on which I needed to focus to achieve the most important results: safety, customer satisfaction and integrity. I had learned from my father that it was better to focus on a few things and do them exceptionally well, than to do many things and perform below par. One cannot boil an ocean or solve world hunger.

Working as a team, we planned to turn the storeroom around. We wanted it to become a place where staff would want to work, that our field customers would speak highly of, and somewhere leaders could proudly visit and show off to external visitors.

I discussed this with my manager and reviewed the plan and expected impact. He felt it was challenging but he was supportive and encouraged me to proceed. His support was key, as the plan would need financing as well as the support of other departments and functions.

The vision I shared with the team was for our stores to be as presentable and organized as world-class supermarkets (Sainsburys, Walmart). I had seen pictures of these stores in magazines, which showed materials properly arranged on shelves, and stores with proper lighting, flooring and enclosures.

We set to work, focusing first on how to improve the security of materials in the storeroom. I worked with engineering to design the required construction work. From the design, they were able to provide costs using SPDC's estimating processes. Then, with the information and business case, I secured management approval. I had to go back on a few occasions to provide data and responses to queries. I then worked with Finance to secure the budget during the next business planning cycle. It was approved and in the new year we started to build walls around the store to secure the materials.

There were challenges during the implementation. We had to phase funding over two years as Finance could not provide all the required amount in one financial year. The contracting cycle was delayed and we had to change strategy mid-way from procuring new materials to utilizing in-house materials (e.g. pipes) in the warehouse for the construction of heavy-duty racks needed for storage of heavy items (e.g. flanges). There were delays from Engineering, as design work, the bill of materials calculation and costing took longer than scheduled. My team held weekly review meetings. We were unrelenting in following up activities, in line with the schedule.

A cold room for sensitive materials (i.e. instruments) was built and we procured carousel machines, synthetic racks and pallets. We had seen these used in those world-class supermarkets that we envisioned. We then arranged the materials. We labelled each and located them in assigned racks and in the carousel units. We entered the location information for each type of material into the company's enterprise system to make retrievals easier. Many of the staff worked extra hours and on weekends to achieve this objective.

Our second focus was customer satisfaction. Material collection required customers to come into the storeroom with a requisition; our staff then spent hours searching for the requisitioned items in the 'junkyard'. When items were found, they were handed over to the customer and the system was updated. This meant our field staff were spending hours away from their core production work, travelling and waiting to collect the materials required for their work.

Working with IT, we were able to enable customers to send their requisitions electronically, without having to physically come to the warehouse with hard copies. We got management approval for vans and trucks to be reassigned to us from the Transport department. These enabled us to deliver items to customers at their worksites. We called these 'milk run' deliveries. Customers no longer had to leave their core work to travel long distances for materials. They could focus on increasing production.

As a result, the warehouse was no longer crowded with customers seeking to collect items. We were consolidating requests and delivering to them instead. This changed their perception and satisfaction levels rose.

These efforts also led to reducing theft to zero. The safety risks were also reduced as the several drivers of customers no longer needed to drive from their various locations to the warehouse to collect materials. We were consolidating all the journeys and, with synergies, the driving exposures were significantly reduced. In addition, injuries were reduced in the warehouse and productivity improved.

At the end of two years, the warehouse had been completely transformed. Customers shared stories of the change and impact with their supervisors and managers in SPDC. As a result, we were visited by the company's Divisional Director and other management (from Engineering and Supply Chain).

On one visit, the Director (Andrew) was shown pictures of the store before the improvements. I watched his face as he looked in disbelief at the transformation that had occurred. At carousel units, Andrew was visibly impressed as he tested the working of the units. He stated that in the years he had worked across Shell's ventures, he had not seen many countries with such facilities. He later cut the ceremonial ribbon and spoke to storeroom staff. He appreciated our efforts and results, and encouraged the team to continue to improve.

I was happy and excited about the results we had achieved. I was no longer going to the supervisors' daily meeting to report on theft. I felt proud of the team for the improved working environment, which also motivated them to improve their productivity. I was invited to share our story in several fora in the company, at general staff briefings and in magazines. It was a proud moment for all of us and we held a celebration party for all staff and their spouses to showcase the achievements.

Periods of improvement works could also have an impact on the social life of staff, as they may be working longer hours. It is important for a leader to encourage staff to look out for one another and identify any signs of stress so they can be addressed. Many staff were married, and some got married during the improvement work. Ensuring family balance while having to wake up early and return late at night due to the drive for improvement requires a spouse who is understanding. I encouraged staff to share our vision with their spouses and to discuss how we were progressing. This helped gain their support while providing us an independent sounding board. I also always had a jotter by my bedside where, when new thoughts and ideas came to my head – even in the middle of the night – I quickly jotted them down. I then discussed these with the team the next day and we decided whether it was appropriate to incorporate them. It was an intense period, but the progress we were making made it worthwhile. To achieve a goal, one must know exactly what the goal is and focus, then enlist critical stakeholders. It is not possible to hit a target with your eyes closed.

Babs Omotowa (1993)

This was the type of transformation that regularly occurred in the roles to which I was assigned during my time with Shell. Such impacts were instrumental in gaining the confidence of my bosses, both expatriate and local. Many of them viewed me as a reliable and transformational member of staff.

I learned that no matter how lowly the work that one is assigned might be, one should embrace it and work diligently. It is important to work to leave the role at the end in an improved state from when one started. I learned that hurdles are to be overcome; when one meets hurdles, one should learn to smile and embrace them, as beyond lies a prize.[5]

Going beyond the frontiers

Other assignments followed during the six years I worked in Warri. These included roles as senior planner and buyer. The planning role was relatively new and without direct bottom-line impact. It was an influencing role and entailed developing business plans for the next year and reporting progress. This included forecasting and exploring opportunities for the medium and longer term.

The role was viewed as an administrative assignment and not coveted by staff. Like the warehouse experience, I set out to understand the unique challenges and opportunities of the role. I became familiar with the routine elements (plans, reports) and ensured that these were delivered to the highest quality. But given that this was a relatively new position, I felt there was room to define and expand its level of impact.

The role provided access to senior leaders – something that could be leveraged, even beyond the company. In one example, I worked with the procurement and logistics manager, Dunni Ososanwo, to set up an industry forum to bring other oil companies in the region together to review working on opportunities of synergies. The scope included how to better utilize excess project materials. Most of the companies had lots of materials and

---

[5] To my team members – Alfred Ogoluwa, Patrick Edobor, Wilson Atumah, Jacob Adewoyin, Peter Akpatason, Gabriel Akporuku and Gordon Duku – who worked with me in the storeroom: my gratitude for their sacrifices which were instrumental the to results achieved.

equipment (e.g. pipes) that were left behind after projects. This was due to traditional procurement of contingency materials, because of the long and complicated supply chain to get materials into Nigeria. Most projects did not want the risk of delays in waiting for materials, so they often bought excess materials. Most companies had tens of millions of dollars of such materials lying idle for years. Others could utilize these, but the companies were all working in silos. The forum explored the opportunity to use these materials rather than allow them to go to waste.

I was excited to help arrange the forum. I contacted managers of other companies and set up meetings to discuss strategies. The forum was successful, and millions of dollars of materials were traded by members rather than allowing them to go to waste or waiting for materials. Progress was also made in sharing facilities and jointly lobbying the government on regulation.

I later took these ideas beyond Nigeria. I contacted managers in other African countries where Shell operated (Gabon, Egypt). We began an 'African Co-Ordination Forum' and held a meeting in Rabi, Gabon in 1996. We explored opportunities including consolidating our material shipments from Europe and sharing excess stocks that existed in each country.

I thoroughly enjoyed the job and expanded it beyond the historical routine administrative elements. By the end of my assignment, it had become a sought-after role.

My achievements were enabled by others, like my boss (Dunni), whose industry drive became an inflection point for me on leadership. It expanded my perspective on relationships and working externally, beyond an organization, to build win–win outcomes that delivered larger synergies.

I learned that significant opportunities can be achieved by working beyond company boundaries, with other organizations that may traditionally be considered competitors, while complying with anti-trust rules. The wider one looks beyond the horizon, the more opportunities can be found. I learned that it is what one *makes* of a job that matters. There is a need to look for opportunities to transform a role. I learned that the attitude one takes to a job matters.

Attitude matters

A process the company had for performance and staff career management was an annual appraisal cycle. It consisted of two key elements: a performance factor (IPF) and estimated potential (CEP).[6]

IPF was the rating of performance, based on set annual targets. It had five categories: Poor; Below; Satisfactory; Good; and High. CEP, on the other hand, was an assessment of the potential (position) that one could reach at the end of a career. It consisted of numerical and alphabetical ratings. CEP was equivalent to the company's seniority structure. IPF and CEP were assessed by supervisors and reviewed by panels.

As a result of the appraisal, we would each be assigned an IPF and CEP yearly. The IPF would be used to determine the annual bonus to be paid to a staff member. CEP would be used for planning: it would play a part in how rapidly staff were promoted, and the types of leadership courses and overseas programmes they could attend.

In our first year, many of my colleagues were dissatisfied with their IPF and CEP, and showed their displeasure at every opportunity. There were some strained relationships with their supervisors, who they believed had ranked them unfairly. This affected the confidence of many of my colleagues and their view of the objectivity of the process. Most of us had similar CEPs (2 i.e. departmental head) at that time.

While appraisal can be subjective and controversial, I observed that their approach was not leading to a different outcome. I took a different approach as I had noticed that, irrespective of the ranking, the company paid a bonus to all staff – albeit differentiated. The bonus paid to me at the end of the first year, despite not attaining a high ranking, was more than my entire salary as a teacher, or that at NSPMC, a few years earlier.

---

[6] See https://elijahconsulting.com/cep-currently-estimated-potential-or-character-expression-potential

My attitude was to simply accept the outcome, and ask my bosses about the areas in which I needed to improve. Apart from the generous bonuses being paid, I believed the best response would be to redouble my efforts. I had learned early in life, during my secondary and university years, that it is important to focus on what you can control and not spend time on what you cannot.

I had learnt that a management system that relied on human interaction (like appraisals) could never be perfect. It was clear that I could not control how my supervisors would rank me; however, I could control how I worked, the results that I achieved and how I behaved. By putting energy and effort into those things, I might even impact the way my superiors might view and rank me.

This attitude meant that, from the start of January of every year, I was already thinking about what I would do and how, rather than spending months brooding and protesting as some colleagues were doing. It enabled me to achieve results, maintain healthy relationships and stand out at that early stage of my career. I learned that attitude matters.

Chapter 3

Ambition

It's only after you've stepped outside your comfort zone that you begin to change, grow, and transform.
– Roy T. Bennett (1939–2014)

Growth through personal development

We're delighted to welcome Babs as the next CIPS President. He will be an asset to the profession and to the Institute as he brings a wealth of experience and knowledge about procurement and supply management as well as business principle. His theme, 'Raise Your Game, Raise Your Voice', will encourage professionals to move out of their comfort zone and translate the value that procurement can offer business in a way CEOs and boards can understand. He is also encouraging professionals to tackle society's issues, including corruption, the environment and sustainability, and to develop talented individuals to strengthen the profession.

This was the announcement by UK Chartered Institute of Procurement and Supply (CIPS)[1] on 27 October 2014. I was humbled. This was not a recognition I had coveted. I had focused on developing myself over the years to achieve competence in the new area I entered when I joined SPDC. I had been convinced that to have a great career in an organization, one needed to keep developing throughout one's career. This was a lesson I learned while teaching and from my parents.

---

[1] See www.cips.org

In 1978, my younger brother and I accompanied our parents to Murtala Airport in Lagos as my mother departed for the United States following a tearful goodbye. She was on her way to New York University to complete a Masters degree in art education. My father had encouraged her to enrol, despite her apprehension about leaving her four boys behind. He said to her, 'Go on and achieve greater heights. I will look after the boys.' On her arrival in New York, she developed bouts of headaches and was diagnosed with stress from missing her family. She soon adjusted, though, and was able to concentrate on her studies. She completed her programme in 1980.

My mother continued her development and earned a PhD when she was nearly 50 years of age. I grew up seeing her continuing her education to reach the highest level in her field, and its translation into advancement at her workplace. It reinforced the lesson of needing to continue to develop oneself as a key enabler of growth.

Personal development is the process of expanding one's full potential. It is the path to growing one's strength, reducing weaknesses and expanding one's capability and knowledge. It is a way to become the best that one can be. It is not possible to stand still with development: if you stop developing, you start to decline in relevance. It is a journey, not an end or destination.

I learned to seek every opportunity to develop myself throughout my career, and to aim to reach the peak of my profession. This approach to development was key to my career advancement.

Rising to the peak of one's profession

With a chemistry degree, working in a materials management role – that is, procurement (materials and equipment) and logistics (air, marine and land transport) – was new terrain, although I had some insight from my MBA thesis on inventory management. I was initially able to learn about procurement and logistics from onboarding and mentoring by superiors. However, it was soon clear that in order to have an impactful career, I needed practical and theoretical knowledge.

CIPS was a leading global certifying body in the procurement and supply management field. As providence would have it, SPDC had set up a training programme for my colleagues and me. A consultant was contracted to help us acquire knowledge in the field and prepare us for professional certification with the CIPS.

The distance learning programme commenced in 1994 and we combined it with work. In addition, some of my colleagues were also starting new families. This really tested us – especially the ability to excel at our daily work and pass the examinations. The experience I had gained at BSMC in multi-tasking (teacher, games master, examination coordinator) was valuable during this period. By 1997, we had completed the CIPS Diploma, which enabled us to improve our overall performance at work.

CIPS published a monthly magazine containing new ideas and case studies from other industries and countries. The Institute also held conferences and branch events to keep members abreast of developments and best practices in the profession.

In 1998, the oil industry went through a price slump. It was not unusual, as the industry had witnessed price changes since 1970. The oil price per barrel, which was US$20 in 1973, increased to US$50 in 1974 and reached US$125 by 1980. It then began to fall, reaching US$20 by 1998. It then rose and reached US$145 by 2008, but again fell to US$30 by 2016. The industry is characterized by such cyclical price fluctuations.

The price of oil, like commodity prices, is determined by supply and demand forces. Supply is the volume produced by oil companies and determined by investment levels and market forces (e.g. OPEC). Demand is impacted by global economic outlook (e.g. manufacturing, travel), population growth (China, India) and geopolitical activities (e.g. wars such as the US invasion of Iraq). High oil prices enable oil companies to invest and produce more, but this leads to reduced consumption and invariably falling prices. Low oil prices, on the other hand, lead to an increase in customer demand, which spurs economic activities but leads to a lower appetite of oil companies to invest; this in turn leads to lower volumes and thus higher prices. This is the basis of oil price cycles, with some suggesting that a more sustainable oil price level is the mid-range (i.e. US$60–70).

In 1998, SPDC's revenue had fallen as a result of the low oil price. In response, the company decided to reduce costs. Training and conferences were cut. The previously approved budget for our attendance at the CIPS conference that year in the United Kingdom was cancelled, disappointing many of us. Some blamed our manager for giving a low priority to training and cutting the budget, but I took a different view, mindful of the challenges the company was facing. Revenue had reduced significantly, and the company had to take action so as not to incur losses.

My view was that development could be done personally and that since we were being paid well, many of us could afford to sponsor our own development. I had learned from my father's self-sponsorship at university from his meagre income working on farms that one always needs to prioritize continued education, even if it requires self-sponsorship.

I went ahead and paid for the conference fees, flights and accommodation. This happened at various times when the oil prices fell. In each instance, I was happy to invest in my own continued learning. I was able to network and interact with colleagues from all over the world and from different industries, and I was able to learn from them about new ideas and innovations in procurement best practices.

By 2000, I had built credibility within the CIPS. I was appointed an examiner for the foundation level, further enhancing my knowledge and network. The role brought nostalgic memories of my teaching days and enabled me to keep abreast of new developments in the procurement and supply profession. I found preparation sessions on marking and discussions with fellow examiners valuable. These sessions ensured consistency in the standard of marking by various examiners. They were great networking opportunities.

Later in 2000, I enrolled at the University of Leicester in the United Kingdom to study for a second Masters degree in business administration, specializing in supply chain management. The programme enabled me to gain knowledge and build new networks with people from other industries. I gained new insights and had opportunities to test some of my own ideas with others. I was able to combine this study as distance learning (based in Aberdeen, Scotland) with work and family life, and graduated in 2002.

I was also active in the setting up of a CIPS branch in Nigeria, consisting of alumni of the Institute, students undertaking examinations and aspiring students. This provided me opportunities to speak at many fora and conferences. I was also active in the wider African branch where I spent time working with officials at the headquarters in South Africa, subsequently being appointed to its board of trustees. I became a regular speaker at many of the African branch conferences in Ghana, Zambia and South Africa.

Knowledge from CIPS and my Masters degree equipped me to do things differently at work. For example, we introduced a strategic 'corner-shop' approach for fast-moving, low-value items. Through competitive tender, we identified a few vendors to supply the items. We then established a process to enable staff to phone directly for items to be delivered or to pick them up, without the traditional administrative controls inherent in our procurement process. The concept was acquired from best practices. Continuous learning is important for improvements and valuable for bringing new and tested ideas that can impact an organization.

Several years later in 2013, during a CIPS conference where I delivered a speech, the then CEO of CIPS, the late David Noble, said to me, 'The board of CIPS is interested in appointing you as the next Global President of the UK Institute. What do you think?'

I was taken aback as I had not shown any interest in the role. It was unusual for one who had not sought the presidency to be approached. Usually the position had gone to a previous Vice-President. I said to David, 'Please give me time to think about this and discuss it with my family.' He suggested that when I was next in the United Kingdom, he would set up a meeting.

During my next visit to London, I met four CIPS board members over dinner and we had a fruitful and engaging evening. We talked about the profession's growing value as well as some of the challenges, including ethics, sustainability and digitalization. We discussed the future of the Institute. I was convinced by the progressive thinking of the board members that this would be a role I would thoroughly enjoy. I also felt it would be one in which I could inspire younger members and those from developing countries.

David met me during my next visit. I had gone to London for eye surgery. I remember opening the door to him the evening after the operation. The shock on his face was obvious when he saw my left eye heavily bandaged. We both laughed as I tried to engage him, with one eye closed, as he shared details of the board meeting confirming my appointment as the next President.

I was CIPS President from 2014 to 2015 with a theme of 'Raise Your Game, Raise Your Voice'.[2] It was a rallying cry for members to become strategic at work and maximize value rather than engaging in siloed thinking. I encouraged members to speak up and celebrate the value they bring to business and society. I travelled to many places (Dubai, Abu Dhabi, Australia, Ghana, Zambia, South Africa, the United Kingdom, etc.) to attend conferences and branch events, to listen and share the thinking behind my theme.[3]

CIPS Global President addressing graduation
students at Birmingham (2014)

---

[2] See www.cips.org/who-we-are/news/new-cips-president---procurement-and-supply-comes-of-age---raise-your-game-raise-your-voice
[3] See www.cips.org/who-we-are/news/farewell-on-my-cips-presidential-year--babs-omotowa

Shell was supportive of my CIPS Presidency and allowed me time off work to travel to the various events and conferences. The company sponsored some of the CIPS events, while I paid for my own expenses. I had insisted that CIPS should not pay for my expenses during my presidency, as it would create a conflict of interest since CIPS also provided contracted training services to Shell. The role brought recognition for Shell within the Institute and with members all over the world. It brought recognition for me too, well beyond Shell.

I learned that one must be willing to invest in their own development and not wait for an employer. I also learnt that continuous learning enables one to bring transformation to an organization and to make an impact, as well as bringing wider recognition.

I also attended leadership training programmes at the International Institute for Management Development (IMD) in Switzerland, the Institut Européen d'Administration des Affaires (INSEAD) in France and Harvard University in the United States.

In addition, I spoke at conferences at Harvard and MIT Universities in the United States. At Harvard I spoke on 'Nigeria: In Need of Dreamers',[4] highlighting the realities, myths, challenges and can-do spirit of Nigeria. I spoke on Nigeria's resources (human and minerals) that make its emergence as a major market economy inevitable. I shared that Nigeria's path to transforming itself from a developing to a developed nation requires dreamers of a certain kind, nation builders with vision, integrity and the drive to enrol talent to make a difference.[5]

I suggested that Nigeria could learn from nations that had moved from the developing country phase to become developed countries. Learning from others is a great way to develop.

---

[4] Seewww.slideshare.net/NigeriaLNG/babs-omotowa-havard-university-speech

[5] See www.vanguardngr.com/2014/05/nigeria-eyes-babs-omotowa

At Harvard University delivering a lecture (2014)

Learning in a developed environment

Since independence in 1960, Nigeria has been a developing country, grappling with social, economic and infrastructural challenges. The subsidiaries and ventures of international companies based there must operate within those realities, despite their global capabilities. One approach used by multinationals to train their staff and transfer best practices into the country is to second staff to their overseas subsidiaries and ventures, so they can be exposed to, learn and bring back best practices.

In 1999, I secured an overseas assignment to Shell UK in Aberdeen, Scotland. The process for overseas assignment was for its joint venture companies to identify talented staff (based on performance and potential) at the beginning of the year and permit them to apply for vacancies in other Shell entities abroad. I had been identified to apply for roles in 1999.

One of my mentors advised me not to accept the assignment. He based this on past experiences of staff who had gone there, describing how many had become disillusioned. He confidently predicted that during the assignment the United Kingdom would rank my performance low and downgrade my potential. He said, 'Do not expect promotion in the UK. No Nigerian, to my knowledge, had ever been promoted there.' He believed discrimination was rife in Aberdeen and that there was a prevailing bias to appraise their own staff better than Nigerians. He felt that cultural differences and conservative attitudes played a part in this.

He advised that I should rather seek an assignment in Oman, Brunei or Malaysia, which he believed were more favourable to Nigerians. If that failed, then he advised I should simply continue working in Warri.

Living in Warri, I faced an altogether different challenge. Warri was a riverine area and public hygiene was poor. In addition, there were always electricity cuts and we had to leave the windows open at night. Consequently, we were regularly bitten by mosquitoes and I often fell ill with malaria. I decided that, if only to escape this, I would prefer an overseas assignment. I decided to take the risk. I believed I would learn more during an assignment in the United Kingdom than in the other countries and that the gain in knowledge would compensate for any career challenges.

I proceeded to the United Kingdom in January 2000 for the assignment in the Aberdeen office. I had left Nigeria with my wife, Helen, and our six-month-old son, Tobi. I had first met Helen while working in Lagos, when she was introduced by her friend who was a relative of mine. She was a student at that time and we were married many years later, after she had graduated, and had our first child, Tobi, in June 1999 at Warri.

The temperature in Lagos when we left was 23°C. On our arrival in Aberdeen, it was –5°C. I remember the doors of the aircraft opening to the cold winter morning, and as Helen began to alight from the airplane she looked at me. She did not say a word, but I sensed she was quietly asking me whether I was sure about what I was doing. However, we ended up living happily in Aberdeen for nearly a decade.

Shell's operations in Aberdeen supports oil and gas production in the United Kingdom. Shell had offshore production platforms in Brent, and in the Northern and Central Areas of the North Sea. To reach some of these platforms required a two-hour helicopter flight or 24–28 hours of sailing. The offshore platforms consist of accommodation and offices for up to 100 staff. They also contain production and storage facilities where oil and gas are produced and processed after extraction from oil deposits 600–900 metres below sea level.

I started in the Business Improvement Unit, responsible for identifying and implementing new ideas to enable the company to earn more revenue and reduce costs. Expatriates have the responsibility of improving company results and developing local staff. A Nigerian expat in the United Kingdom was a rarity then, especially as Nigeria lacked many best practices and we were unlikely to be able to improve practices in a developed country. The reality was that Shell considered expat assignment as a developmental experience. For those from developing countries, it was an opportunity to learn, and to take knowledge back to our countries after our assignments.

I settled in quickly and made new friends. One was a Scotsman named Tim. He was knowledgeable, experienced and friendly, and he warmly welcomed me to the team. His wife was Lebanese and he had travelled to Lebanon but had never worked outside the United Kingdom. He was keen to understand the culture and life of Nigeria. We arranged social events where our families interacted. Summer was short in Aberdeen and on the rare occasion that the weather was sunny, we would barbeque, with Tim usually at the grill. This type of work friendship that extended to the families was invaluable, particularly as spouses of expatriate staff often can't find work immediately on their arrival in a new country, so initially have few friends and limited interactions. Being able to meet spouses of colleagues made settling in easier.

I took time in the first few months to read and understand the processes. I was willing to be vulnerable in public when I did not know things and to ask colleagues why things were done the way they were. Tim was always helpful. I also did not shy away from making suggestions, no matter how basic or unsophisticated they might seem.

Once I suggested the use of hovercrafts as a replacement for helicopters used to transport staff to offshore locations, which cost the company several million dollars per year. The suggestion was met with laughter and became the butt of jokes. I had not carefully considered the rough water and windy conditions of the North Sea. However, I continued to offer new suggestions, convinced that unorthodox ideas were necessary to make an impact and would lead to positive outcomes for the company.

Later, I made an unorthodox suggestion that *did* make an impact. The company spent tens of millions of dollars on offshore supply vessels that supported offshore production, projects and drilling rigs. These vessels carried on their decks containers filled with material and equipment. They also had liquid tanks underneath that were used for transferring fuel and chemicals.

Vessels were being utilized separately for drilling and production. We had many vessels going to the same area, some half-empty, because of the dedicated nature of their operation. I suggested breaking these silos and combining the supply vessels to both our drilling rigs and production facilities. The combination was to be based on geographical locations, rather than on a functional basis. This meant a vessel would support not only a rig but also the production facility in the same area.

Modelling the business case, I was able to demonstrate that utilization of a vessel could be improved from 60% to 85%. Management liked and approved the idea, but not without resistance. I recall that during the pilot, the Head of Drilling, who was accountable for rigs across the United Kingdom, told me it was an ill-conceived idea and would fail. He thought that while there may be savings in vessels from the initiative, it introduced risks; any downtime on a rig waiting on material would quickly wipe out the saving. At that time, the daily rate of an offshore rig cost nearly 10 times that of a vessel. As a result, if a rig was unable to work due to non-availability of materials, the impact of the loss of a day (idle) of the rig was significantly higher than savings from a vessel.

I reflected further. I rechecked the model and the business case and discussed it with other offshore staff, including those on the rigs. I was convinced that we had thought it through and simulated the various scenarios. We had back-up plans for the new arrangement (e.g. emergency vessel call off agreements). I was willing to go ahead and take the risk.

We completed the pilot and it worked as planned. We then proceeded to full implementation and saved several millions of dollars by cancelling over 30% of the hired vessels. We managed the operations and there was no downtime to any of the rigs. Years later, the Head of Drilling acknowledged that the initiative had achieved significant benefits for his operation. It had improved flexibility and the rigs had more vessel options than they had previously.

I learned to model ideas, identify risks and design solutions, knowing that not all bosses would be easy to impress as some had different drivers and expectations. I learnt that once all options and information are considered and discussed with experienced players, one needs to go forward and be prepared to accept responsibility. In moving forward with solutions that are not life-threatening, one must think 80:20 and not spend a disproportionate amount of time trying to get solution to 100%. In driving change, one must not be stuck in unending analysis paralysis, like a rabbit caught in the headlights.

I took the idea beyond Shell's operations. There were other oil companies (e.g. BP) with rigs and production facilities around our area of operation. They were also sending vessels to their locations in the same area. With time, many of the other oil companies also adopted the same model. We were also able to buy space on each other's vessels, which enabled us to avoid taking on additional vessels and together saved tens of millions of dollars. The flexibility of response to all the rigs and platforms in the North Sea improved. The offshore customer satisfaction went up and we made a positive impact on the bottom line of all the participating companies.

The North Sea extends to other countries including Norway and Holland. Shell had operations in those countries, so I saw similar potential to share vessels across all North Sea countries where Shell operated. I reached out to colleagues in those countries and we set up an informal group of marine managers. We pulled information on our vessels together and saw opportunities to share vessels (anchor handlers, supply vessels). We did this slightly differently; the geographies were not as feasible for combining supplies due to longer distances. We decided to move vessels around countries in rotation by periods, basing demand on planning and forecast.

We were able to achieve significant savings of tens of millions of dollars for the Shell companies involved.[6]

Shortly after we had implemented this regional vessel sharing, Shell separately undertook a review of its entire organizational model. The company elected to move from a country-focused organizational model to a regional (multi-country) model. Staff were sceptical about the practicality of the design, but management soon identified the regional vessel sharing as demonstrating the practicality of a regional organizational model. It was quickly used to showcase that regional approaches offered significant opportunities. The recognition of the pacesetting vision and transfer of learning across several companies and countries was extremely satisfying.

Later, I raised the application of the resource-sharing idea that had been implemented in the United Kingdom and Europe to our Shell offshore operations in Nigeria. It would have the same positive impact with significant savings. This reinforced the expectation of Shell that expatriation from developing countries was a way to transfer best practices.

These types of outputs also led to personal rewards. I attained managerial status within a decade of joining the company, having been promoted nearly every two years. This was mainly due to the creativity and innovations that I brought into every role, the results and impact that were achieved and the working relationships with others, both within and beyond the company. Progression is related more to results achieved and impact made, rather than the number of years spent in a company.

I learned about maximizing the value of a successful idea by replicating and transferring the idea beyond an organization and country. I learned that while many ideas may not materialize, one must continue to generate new ideas as the process helps to improve quality, and at some point one of the ideas will make a difference. I learned not to be shy but to speak up and be comfortable stepping beyond my comfort zone.

---

[6] To my team members – Tom Macdonald, Hans Van Rossum and Erik Gjul – my appreciation for their dedication, which enabled the delivery of results.

Going beyond the comfort zone

By 2005, it had been over a decade since I had joined SPDC. Since those early days, I had felt it was important to work at the core part of the business (e.g. production). I had developed expertise in my primary discipline of procurement and supply, and was in my comfort zone. I felt that remaining on the sidelines of the core area would limit my growth. I needed to gain experience of how the company made and spent money. This would enable me to learn and make me more rounded. As a result, I regularly sought opportunities. Such a move would take me beyond my comfort zone, but it would enable me to learn. Remaining in a comfort zone breeds complacency.

Thus, while in the United Kingdom, I was keen to explore opportunities to work in those front-end departments. For someone who started in the storeroom, it was always going to be a challenge to be allowed to work in such technical capacities, especially at the relatively senior level that I had attained within a short time.

An opportunity came when Shell UK partnered with IBM and introduced a new improvement methodology (Lean Six-Sigma) that was developed by Motorola (USA) in 1986. Based on the Lean manufacturing business model created by Japan, it relies on a collaborative effort to improve performance by systematically analyzing, identifying and removing waste, improving quality, and reducing variations and defects. Ultimately, it leads to higher profits for an organization. During my CIPS study, I had read how it was instrumental for Japanese manufacturers and car industries to compete globally. Using the methodology, they were able to achieve 100% quality, eliminate rework and rejects, and significantly reduce costs. I had been a fan since then.

I applied to be trained, and spent months learning Lean Six-Sigma. Subsequently I was assigned to apply the learning to improve our production processes. This was in the area of maintaining equipment in the oil production field. With this, I was able to gain deeper understanding of the company's core processes and we re-engineered a new process (Campaign Maintenance) that delivered significant results across the United Kingdom.

With the impact and credibility from this experience, I was appointed Production Superintendent for the Pierce FPSO (Floating Production, Storage and Offload) facility in the United Kingdom. FPSO is a floating vessel that receives hydrocarbon from sub-sea installations thousands of feet below sea level. It then processes, produces, stores and exports oil and gas via tankers. It can be a converted tanker or purpose built, and the FPSO floats – unlike an oil platform, which has fixed legs to the seabed, although both serve similar purposes. The FPSO is connected to the seabed through a system that consists of a turret integrated into a vessel and fixed to the seabed by means of a mooring system. The turret contains a bearing system that allows the vessel to rotate.

FPSO

This experience in a core production role was what I had always wanted, and to be given the chance by my bosses in the United Kingdom was a sign of their confidence in me. It was a unique learning opportunity to understand the core operations of the business.

The new role gave me opportunity to work closely with offshore staff and to demonstrate leadership. I had initially learned leadership from reading books but more from studying my direct bosses and company leaders. I noted which leaders got the organization motivated towards company goals and achieved results. I had experienced that leadership is situational, but mainly about inspiring people to achieve exceptional results. I realized that building emotional connection and care is important to aligning people to give their best towards achieving a vision. It also requires giving them the space to deliver or to make mistakes, but being there to support, develop, coach, mentor and recognize them.

In my seven years working in Shell UK, I did not experience discrimination or prejudice, contrary to what my mentor had warned me about in 2000. It may have existed subtly, but I did not spend any time looking for it or worrying about it.

My experience was that the system that Shell had for appraisal was not finalized by individuals but by a panel, and due to the company's diverse workforce, such panels were usually diverse. The diversity in leadership and policies on inclusion reduced the chances of institutional discrimination.

I also believed it was an example of something over which I had no control, so I spent my time focusing on how to deliver results in my assigned tasks and on improving things. By achieving the objectives, the notion of being discriminated against and poorly appraised never materialized. Rather, I was rated highly and had a positive experience in the United Kingdom. I received multiple promotions and my potential was elevated to the highest level in Shell (Senior Executive Group i.e. top ~100 leaders in a company of 100,000 staff).

I learnt that stepping outside one's comfort zone brings significant benefits to expand one's skills and gain broader knowledge. I learnt that a positive attitude to continuous learning and applying the knowledge to improve work makes a major difference in career advancement and professional progress. If one aspires to a different future, one has to invest today, change what one is doing and go beyond their comfort zone.

To change others, start from self

With the start of a new regional organization in Europe in 2003, I was appointed a Manager in the United Kingdom. I also had a new boss, Carl, an intelligent, high-flying engineer.

Carl and I had a challenging start. On many occasions he showed disdain for my work and made negative comments. I had not experienced this with previous bosses. The relationship deteriorated rapidly; it seemed the more effort I put into my role, the worse the assessment from Carl was. Working became quite stressful. I felt I was being unnecessarily put down for reasons that I could not understand. It reminded me of the comments my mentor in Nigeria had made a few years ago. I questioned whether Carl was being discriminatory due to racial prejudice, as I noticed that he had a totally different disposition to my other Caucasian colleagues.

Arriving home from work one evening totally frustrated and disillusioned, I was convinced that I had reached my limit. I informed my wife that I was going to resign from the company the next day due to the continued stress arising from a poor working relationship with my boss. I could not take it anymore.

I went into the study and wrote my resignation. I thanked the company for the opportunities it had provided. I explained that the reasons for my resignation were personal and wished the company well. I addressed the envelope to the HR Manager and put it into my office bag. I felt relieved, as if a burden had been lifted from my shoulders.

My wife watched silently. Over dinner we discussed our younger son, Fiyin's, kindergarten schooling. He had only just started and was still adjusting to the new experience. He was, however, making new friends and seemed to relish the experience. Just before we went to sleep, my wife said she had questions about my plan to resign. 'Did you join the company because of Carl?' She asked. I said no. 'Why then do you want to resign because of him?' I had no answer. She said, 'I have nothing against you resigning, but it should be on your own terms. I can understand if you have a different aspiration and want to do something new.' Her comments continued to play in my head throughout the night. By morning, I concluded that she was right. I retrieved the resignation letter and tore it up.

Her subtle intervention was an example of how important family support is, especially during overseas assignments. Her presence as a partner, counsel and supporter during challenging times was immeasurable. Being away from one's country and local employing company subsidiary can sometimes feel lonely at work. The cultural differences and sometimes internal politics that play out in overseas subsidiaries or company headquarters are examples of why having a partner who understands one's context and can take a more dispassionate view is so helpful.

Over the next few days, I reflected on possible reasons for my poor relationship with Carl. I remembered my father's use of introspection when I scaled the fence one night, which made me look inward at myself deeply. I reasoned that I was the one who had to change the dynamics in the relationship with Carl. It might be naïve to expect anything different from Carl without any change from my end. I guessed that I may have come across as crowding his space, or hadn't recognized him enough as the leader in the team, particularly as he had only just started work in the department whereas I had a few years of experience there.

I recalled noticing a frown on his face during a meeting we attended where I had presented some new ideas. I may have surprised him. I took the decision to change my approach. I was going to ensure that he received respect and recognition. In the past, I had tended to act unilaterally with an authority I was used to, so I started to check things with him before I took any action. I ensured it was Carl rather than me who informed other senior leaders of any new ideas and improvements. For meetings he attended, I provided adequate pre-reads for him on historical context and possible options he could suggest.

Changing the way I related to Carl and the dynamics in the team led to changes in him. He became more relaxed, much more supportive and started to coach me on my career. And he was a terrific coach! I had not realized that earlier. He gave me deeper insights into how I could be more effective. In areas where I was lacking, he sought to set up other coaches for me who were more versed in those areas. These were traits that I would never have known and benefited from had I not changed my own approach, or if I had resigned.

In one of our sessions, I reflected with Carl on how our relationship had turned for the better and he acknowledged that my change of approach and attitude was instrumental. I inquired why he had not raised his concerns directly with me rather than acting it out, and that his indirect approach had stressed me and almost led me to quit the organization. He was surprised, as he had not realized this. He accepted that he could have handled it better but said he was still in the process of settling down in the department. He said he was thankful for my 'upward coaching'.

I had learned that one cannot change others, but one can change oneself. That then improves the possibility that others may decide to change. Irrespective of situation, viewing the other person as the issue is not always a helpful way to get good outcome. One should not spend an inordinate amount of time and energy getting upset, angry or frustrated by other people's thoughts and behaviours.

Introspection is the first step on the way to a fruitful relationship. It is important to be reflective and stand in the other person's shoes. It is also necessary to role model the behaviour one expects from others. If you don't treat others well (supporting, respecting and celebrating them), there should be no expectation that they will treat you well.

My assignment to the United Kingdom was for four years (2000–03), but Shell requested that it be extended for a second term (2004–07). This was unusual at the time, and reflected the high regard in which I was held there. However, in 2006 I returned to Nigeria a year earlier than planned. This was at the insistence of the Head of Shell Africa, Ann – an American. I came to her attention during a staff ranking conducted in Holland head office. The exercise was led by the Upstream Director and included all regional heads of subsidiaries. The Europe Head (Tom Botts) had put my name forward as one of those with the highest potential ranking (Senior Executive Group). It was unusual for the European branch to rate a Nigerian so highly, but they valued the capabilities I had demonstrated, being ahead in regional thinking with the prototype I had developed for sharing marine vessels across Europe. It included the relationships I had developed, including with other oil companies in the United Kingdom and other Shell subsidiaries in Holland and Norway. It also included significant savings across several companies.

I had not met Ann before, but she insisted that if a Nigerian was so positively viewed in Europe, then she needed that individual back in Nigeria to help with the challenges the company faced there. Ann, an intelligent and courageous leader, had joined Shell after working for Mobil, so she was not the conventional Shell staff member. She was direct, tough, unpretentious, straight shooting and with an eye for value. She was well networked and able to confront the significant challenges that Nigeria presented (including government policies and militants). As the first female boss of Shell in Nigeria in over 50 years, Ann presented a culture challenge for some staff in the company, who viewed her strong leadership as intimidating. One thing that I learned and admired most was her humanity and ability to connect with staff at a personal level. She took note of her team members' families and cared about their welfare and progress, and regularly invited spouses to social events. I later worked directly in her Shell Africa leadership team for three years. She was the best boss I ever had, and remains a most appreciated mentor.

Continuous pursuit of personal development through my career was instrumental to my growth. It enabled me to gain more knowledge and greater insights. Through learning, I was able to identify and replicate best practice from other industries, and bring new improvements into the workplace, which helped me to stand out.

Development occurs not just in classrooms but also from working in different roles, new environments, new relationships, professional associations, coaching (on-the-job, mentoring) and networking. It is important to use the opportunities an employer provides, but you must also be willing to invest in your own learning, especially as learning will remain with you beyond your current employer.

This attitude of continuous development was embraced by my wife during my assignment. Like my father, I encouraged her to pursue her dreams. She obtained Bachelors and Masters degrees in social work during our sojourn in the United Kingdom. Spouses of staff on overseas assignment can find it challenging to transfer from a home country employment where they worked, to one in the new location. There may not even be a branch of their firm in the new country. But such overseas assignments provide opportunities for spouses to develop themselves through education.

Chapter 4

Making an impact

*I alone cannot change the world, but I can cast a stone across
the waters to create many ripples.*
                                    – *Mother Teresa (1910–97)*

Going against the grain

'How do you justify your decision to award the contract to this
local company? Are the risks not too high?' Brian asks. He
has only just become my new boss and has invited me to his office.
I wonder why he is interested so early, and sense the Head Office
Aviation team may have contacted him with their concerns.

'Brian, technical experts audited and passed the local company
and their partner as being capable to provide the service before
we included them in the tender. They submitted the most superior
bid by far,' I reply emphatically. 'This is the right decision.' I share
with him the details and analysis behind the decision. He has a
few more questions, which I answer, and I notice that he is taking
copious notes.

As we finish and I stand up to leave, I say to him in a parting
shot, 'I have responded to several queries on this issue. I stand by
the decision and accept responsibility.'

The decision involved the changing out of an international
incumbent of several decades to a local contractor, who was to be
used for the first time in the industry. Brian nodded and said firmly,
'Okay, but Babs, I hope you know you are taking a huge career risk!'

To achieve a successful career, one must take personal risk
to make an impact within and beyond one's area of influence.
One must always be willing to live with the consequences of

any decision. Making an impact means going the extra distance. It is about breaking new ground, spearheading new ideas and envisioning tomorrow's solutions yesterday. It is the confidence to speak up and share ideas and innovation while also listening.

Supporting local industries and employment

Despite Nigeria being the biggest economy in Africa (worth US$500 billion in 2014),[1] its under-development, unemployment (27%)[2] and 90 million citizens living in poverty on just US$1 a day[3] mean it is described as a ticking time-bomb. Its economy is mono-dependent on oil (80% foreign earning, 70% revenue); however, despite the huge spend in the oil industry, it contributes less than 20% of GDP (e.g. employs less than one million people or 0.5% of the population).

One reason for the industry being a low employer is the limited manufacturing that exists in Nigeria. Local goods account for just 10% of the industry's needs and foreign manufacturers have not invested in building manufacturing plants in Nigeria, which would have employed large numbers of locals.

SPDC introduced a local content policy in the 1990s, to increase spending with local vendors and encourage partnerships with overseas manufacturers, and invariably set up local manufacturing. There had been a historical preference in the country for foreign manufactured goods and unfortunately local vendors were more interested in consuming profits than investing. As a result, little progress was made.

The government promulgated a Nigerian Oil and Gas Industry Content Development Act in 2010, to drive local manufacturing and services. However, little progress was made in the early years. There was no master plan developed at the onset and the regulator (the Nigerian Content Development Board) initially focused more

---

[1] See www.reuters.com/article/nigeria-gdp-idUSL6N0MY0LT20140406
[2] See https://tradingeconomics.com/nigeria/unemployment-rate
[3] See www.vanguardngr.com/2019/02/91-million-nigerians-now-live-in-extreme-poverty-world-poverty-clock

on contract approval and compliance than on attracting investors to build local industries.

In 2008, I was appointed Sub-Saharan Africa Vice President, responsible for Infrastructure, Aviation, Marine, Land, Health, Safety, Environment, a role with 600 staff and $500 million in assets. In this role, I saw an opportunity to drive significant impact for both the company and society. With my team, I reviewed many opportunities to improve and develop local capacity, and one of them was in the helicopter category.

SPDC operates in Nigeria offshore (up to 120 kilometres from the coast), and in onshore (land and swamp) locations across the Niger Delta (5.5 million hectares). Due to the long distances, personnel travel to and from these locations by helicopter.

The incumbent helicopter provider was an international firm that had been providing the service for many decades. It demanded a huge rate increase with a bullish 'take it or leave it' approach. We negotiated, but as the company had the monopoly, we had to acquiesce to the significant cost increase. It highlighted our disadvantage in the relationship and the need for us to develop local alternatives.

We assessed the aviation landscape to establish whether there were any local helicopter companies with potential. We encouraged local players to find credible international helicopter operators with which to partner. That way, they would learn and over time develop their own technical capacities to a point where they could compete in the safety-critical area.

Two of them took the challenge seriously and established partnerships with companies in Canada and Denmark. Shell head office aviation technical experts later qualified them as acceptable to operate in our challenging offshore and land locations. This was good news.

We carried out a competitive tender and one of these companies, Caverton, in partnership with Dancopter, won the bid. Caverton had offered a much lower rate over the contract duration than the incumbent.

The incumbent, who had worked with Shell globally, had not expected this outcome. Shell Aviation's head office questioned Caverton's competence, considering the high safety risk and as it was a relatively new and untested player. But it had previously qualified Caverton. This was the background to my meeting with Brian.

Caverton and Dancopter later submitted a transition plan. I noticed they planned to hire six helicopters from overseas companies for the duration of the contract. This meant they would not own any aircraft at the end of the seven-year contract. I viewed this as unhelpful to developing local capacity, as they would end up as mere middlemen. This was the same approach that had been taken by local contractors over the years, which had not led to progress in SPDC's efforts at developing local content. Despite huge spending with local vendors, they simply executed contracts as middlemen, without building any real local capacity.

We discussed why they were not planning to buy their own helicopters. They indicated that they did not have sufficient funds. I asked why they could not approach banks for a loan, armed with the contract from us as collateral. They said they could, but that Nigerian banks' interest rates (20+% per annum then prevailing) made such an option unviable.

This was a dilemma. They could not secure funds to buy helicopters; however, it would also not be appropriate, after a seven-year contract worth hundreds of millions of dollars, that they would not own any assets. I reasoned that as SPDC was committed to developing local capacity, this represented a unique opportunity to act outside the box.

We developed a proposal whereby an interest-free loan (US$85 million)[4] from Shell ventures in Nigeria would be extended to them to acquire six brand-new helicopters. This would be backed by their bank guarantee as security. We would then deduct the loan

---

[4] See http://businessnews.com.ng/2012/03/13/caverton-receives-85-million-facility-from-shell

from their contract invoice payments over the seven years of the contract.

Justification included the safety of our staff – the need for them to be transported by new helicopters instead of leased old aircraft. In addition, such support would be consistent with SPDC's local content policy and enhance the company's reputation as a multinational committed to local development.

The proposal was initially rejected by head office (Finance), due to a concern that with Nigeria being a high-risk country, and as the concept was unprecedented, the proposal and amount were too risky. I clarified how we would minimize risk through the bank guarantee and invoice deduction. The proposal was finally approved in 2011 by Shell and the other joint venture partners. The loan was extended and the Caverton-Dancopter partners went on to acquire the new helicopters.

The transition was challenging. Some staff (pilots, engineers, support staff) and records had to be transferred from the incumbent to Caverton. There were understandable differences in the expectations between them and we had to intervene and deploy some of our resources to assist.

Another challenge was the clearing of imported front-runner helicopters to be used before new helicopters could be manufactured. The agent did not initially provide evidence of customs duty payment. Having learned from the experience with Panalpina (freight company), which was fined US$86 million[5] by the US Justice Department for attempting to bypass official requirements to pay specified custom taxes to government accounts, we took no chances and insisted on the official customs receipts. They paid the fees and obtained the receipts.

The partners had also not yet formed a fully fledged partnership prior to the contract being awarded, and needed to resolve issues such as who held which senior positions; which management systems to use; how much to be paid for maintenance

---

[5] See www.swissinfo.ch/eng/business/panalpina-fined-millions-in-bribery-case/28713696

by the technical partner; how to administer the loan from SPDC; and how to integrate their assets.

On occasion, tempers flared. I had to attend one of their meetings in London, which coincided with another meeting I had in the city. I was clear we had no plan to be an ongoing party in their relationship. We needed them to work together. I said, 'Your local and global reputations are at stake, so approach your relationship on a give-and-take basis.' They later settled on how to run the services. Where appropriate, we offered additional support, such as awarding the contract as a seven-year rather than the original five-year-plus-two-years option. This was to make their loan repayment easier and not put them under excessive pressure.

By 2017, they had provided the service for seven years without any incidents. Their safety record was better than all the other helicopter providers in Nigeria. Their service record (on-time departure, customer satisfaction) was better than historical. Their rates remained competitive without any increases, as we had previously become accustomed to. In contrast, the rate was reduced during a two-year extension (2017–19). More importantly, the loan extended to them was fully repaid without any default, despite the short duration.

Caverton developed capacity, and went on to win tenders for work with other IOCs in Nigeria (Chevron) and other African countries (Total Cameroon). It built a maintenance centre, accredited as the West African hub for aircraft repair overhaul (MRO)[6] and employed hundreds of staff. Caverton's commitment to build a legacy was crucial for the company's sustained success. International development organization took note of the initiative, as a case study for developing countries on how to cultivate local capacity.

I learned that improving local capacity can be done and has significant advantages. It does require thinking differently and having the courage to stand up, even at significant personal risk. Had things gone wrong (e.g. an accident or loan default), there

---

[6] See https://caverton-offshore.com/companies/caverton-maintenance

would have been personal consequences for me. But such risks are worth taking.

Transforming from cost centre to revenue centre

In 2009, another category on which I focused with my team was the development of local capacity in supply bases. Shell Nigeria owned and managed two supply bases in Port Harcourt Kidney Island and Warri.

The bases (warehouse, yard, quays) were used to receive, store and issue materials. Joint operating agreements (JOA) between all the joint venture partners stipulated that contracts above a certain threshold required approval from all the joint venture partners, primarily Nigeria's National Oil Company (NOC). Being a government parastatal, the approval process of the NOC was bureaucratic, took two years, required significant effort and did not always lead to the best outcome.

SPDC contracted vendors to provide human resources, equipment and services (security, maintenance) for the bases, at a cost of tens of millions of dollars. Due to approval delays, the company operated with interim contracts in many instances or put in place multiple short-term contracts to bridge the gap while awaiting formal approvals. This was not sustainable as the inefficiencies were costing a lot and eroding value.

To improve the situation, I shared with my team my experience of how similar bases were operated while I was in the United Kingdom. It would require us to change our approach from 'cost centres' to 'revenue centres'. Such outside-the-box thinking was required to enable us to come up with solutions to address our challenges.

I directed the team to proceed on that basis. We had to find capable third parties to lease our base and manage it. They would attract companies to use the excess capacities that existed on the base and pay for such services. They would then use part of the revenue earned to offset the cost of providing us with the services we required, as well as upgrading the base. They would then credit us with the excess revenue.

We identified Century Energy Services Limited (CESL)[7] for Kidney Island and Temile Ltd[8] for Warri (Ogunu). CESL was an indigenous operator of an offshore oil floating facility. A key incentive for it was the use of our base to support its own operations. Temile was an indigenous Warri contractor.

During the process to put this arrangement in place, many were certain that the NOC and the major co-venturer would not approve it, and would insist on being involved in identifying the suitable third-party company. Such involvement would again lead to bureaucracy, delays and increased risks of introducing potentially incapable entities into the tender.

Given that we would not be spending but rather earning money, I insisted that we did not need approval of the co-venturers for this arrangement. The JOA would mostly require co-venturer involvement in contract approval when we would be spending money (cash calls); however, that was not the case with the proposed arrangement, as we were not going to be spending the joint venture's money.

When this was brought to the attention of SPDC's legal department as part of the contracting process, it advised against proceeding without NOC approval. I thanked the department for the advice, but affirmed that it was an operational decision within the remit of SPDC as the JV operator. I was accountable and willing to take the risk. I signed an undertaking acknowledging that I was acting against legal advice, then I authorized my team to proceed.

The contract commenced and saved us tens of millions of dollars annually for services. For the first time, we actually started earning income from the facilities. We received services at no cost from CESL and Temile. They upgraded the infrastructures

---

[7] See https://ceslintlgroup.com/subsidiaries/century-ports-and-terminals-limited-cptl

[8] See www.temile.com

on the bases. We also did not have our staff chasing after contract approvals from the NOC for years. Those issues were all eliminated along with the ingrained risks involved in the contract approval process.[9]

Two years into the initiative, an unsolicited letter was received from NOC, advising that it had become aware of the initiative and supported it. It was unprecedented for NOC to approve a contract for which its approval had not been sought. I believe this was driven by the value of the initiative to the shareholders. The initiative continued for a decade, without any issues.

The Nigerian government and the regulator need to translate local content policy and laws into concrete deliverables, and the oil industry needs to leverage its high-capital intensive spend to create local capacity and multiplier effects, and avoid the catastrophic backlash that could result from the 'time bomb'.

Industry leaders must be willing to take risks to change the status quo. It requires out-of-the-box thinking, determination and courage to drive changes through, despite challenges. Changes made to the helicopter services and supply bases brought both reputational and financial benefits for the company.

Value can be created when the status quo is challenged and better ways are found. I learned that making an impact requires courage and a willingness to stand alone if required (even against legal advice where appropriate). Others may see challenges, but a leader must find solutions and have a roadmap. I learned that paradigm shifts in thinking come with a multiplier effect on value.

Valuing staff safety and welfare

As a Vice President in Shell Sub-Saharan Africa, the safety of our installations and personnel was part of my responsibilities. A

---

[9] To my team members – Richard Evans, Pius Okediama and Henry Mogbolu – my appreciation for their efforts which were instrumental to the successes on the helicopter and supply base initiatives.

few months before assuming the role, there was a major accident along SPDC's pipeline at the remote town of Iriama in Delta State, south-western Nigeria. Vandals had cut pipes, attached valves and hoses, and siphoned crude oil before they were discovered and stopped.

Our crew carried out repairs to the pipe, including welding. They encountered a gas leak, which led to an explosion and fire, fatally injuring seven workers. This was devastating and affected all staff, as we had lost colleagues and friends.[10] An extensive investigation led to us changing some of our processes and controls. There was a concerted drive to ensure disciplined implementation of company controls (e.g. continuous gas testing, flushing of pipeline with water) as well as further improvements (e.g. double isolations) to ensure accidents like this did not happen again.

On assuming the VP role, with over 10,000 direct and 20,000 indirect staff working in the regional organization, I drove the implementation of the improvements. The steps we took enabled us to operate for over 600 days without any fatalities or serious injuries, until my departure from the role. This was an example of the background of my safety experience and values before I assumed the role of CEO of Nigeria LNG in 2011.

On resumption, I reviewed the company's safety records and was disheartened. It showed NLNG had suffered an average of two fatalities and nine serious injuries every year since inception 12 years before.

Three months into my tenure, we had a fatality. A 26-year-old specialist contractor was draining our slug catcher (the vessel between the pipe outlet and processing equipment, to catch heavy fluids or slugs) using a corrosion probe. The tool suddenly retracted and hit him under the chin. The loss of a life on my watch, was not an experience that I had contemplated, and it was the saddest day of my career.

We investigated and I convened a management review. As the fatality was an unacceptable situation, it was crucial to ensure that

---

[10] See www.reuters.com/article/nigeria-shell-accident-idUSLH2753220081117

there was chronic unease among the leaders. I highlighted that if the trend over the past 12 years continued, there would be 10 fatalities and 45 significant injuries during the five-year term we would be in office. I asked if this was a legacy for which they wanted to be remembered, as it was certainly not one I would accept. Never again should we record another fatality under my watch.

There was unanimous agreement that we wanted a different legacy. We evaluated our leadership and identified what to do differently. One was visibility – I had asked which leaders regularly visited sites; only the Production GM and I had. We agreed that this had to change. All management began to visit sites, not to find faults, but to identify the challenges affecting frontline staff and help resolve them. Visits became occasions when staff also assessed how supportive managers were. They motivated lower level supervisors to do likewise and visit sites more frequently.

An independent review was carried out on our management systems by Shell, as a shareholder with 100 years of experience. The review revealed gaps – blind spots – and identified improvements (e.g. worksite hazard management). We then implemented the improvements as barriers against the gaps.

The focus of management was sustained through weekly meetings. Slippages were addressed and support provided where needed. No other fatality occurred for the remainder of my five-year tenure (2011–16) and we worked for over a year without any major injuries.[11]

I learned that one of the most important roles of a leader is protecting the safety and welfare of staff. This makes staff feel valued, and they are far more motivated when they are convinced that their leader cares for them. A focus on improving safety also leads to improved company performance.

---

[11] I thank all staff in the company including the health, safety and environment team – Shehu Ahmed, Niran Fadeyibi, Kingsley Makasi – for their focus, efforts and dedication. This was my most fulfilling achievement in this role.

Unlocking monopoly and encouraging competition

Another area of my focus upon resumption of work at NLNG was the company's head office. I was unhappy about the conditions under which staff were working at the temporary office in Port Harcourt. The rented offices were not conducive to productivity, and the rental costs were exorbitant.

Port Harcourt lies along Bonny River in Southern Nigeria. It has a population of two million (2006) and hosts many oil companies (Shell, Chevron, Agip). The city greenery and peaceful environment led it to be called the 'Garden City' in 1970s. However, from the late 1990s, the advent of cult-groups, militancy and a high crime rate led to migration of expatriates and the relocation of businesses.

The board had approved relocating the head office from Lagos to Port Harcourt a few months before I resumed. The rationale was to operate in the state where our gas plant was situated. But it also had political undertones, showcasing that businesses were returning to Port Harcourt.

The speed of the relocation necessitated the move into rented offices for an interim period, to enable construction of a head office in the Amadi Creek area of Port Harcourt. NLNG had acquired land in the area a decade earlier, with plans to use it for the construction of a warehouse and a jetty.

Reviewing the land acquisition document, I was unhappy about terms that gave the landowner the exclusive right to carry out all construction activities and to provide all services (water, electricity, sewerage). Based on my experience in contracting, this was not beneficial for NLNG. It was a monopoly, and for a project of this magnitude (a 600-desk office building, jetty and warehouse), this was inappropriate.

In addition, the rental rates we were paying were high. We had to negotiate significant reductions for the renewal period, which we achieved by identifying alternative office in a shareholder's complex. Opening the possibility that we could move away enabled us to achieve reductions.

I set out with my team to untangle the monopoly. I requested the landlord to waive the provisions in the land agreement as a basis for them to participate in the tender. They should still be confident in their ability to win the competitive tender fairly. They refused.

Regaining leverage was key for us to succeed in negotiations. That was how we had achieved reductions in the office rental. To achieve that for the land, we would need to generate a credible alternative location to build the head office. We were not obliged to build anything on this land.

By coincidence, the state government was in same period undertaking a 'Greater Port Harcourt' development (office/residential) near the airport. I approached its management committee to view this layout as an option for our head office. The committee was excited as, being a new development, a high-value client like NLNG would be an ideal anchor client and provide great marketing.

Our interest in the alternative signalled that we were prepared to walk away. Our landlord reached out and I insisted they waive the monopoly. I asked why they were afraid of competition; they had advantage of synergy with their proximity. If they insisted on monopoly, we would proceed with Greater Port Harcourt. Our determination and risk of them not participating in tender was clear. They agreed to waive the monopoly.

As I had been transparent with Greater Port Harcourt's management committee that we had land at Amadi Creek and were only considering land in the new development, they understood when I informed them of our decision to proceed with our existing land.

With the monopoly untangled, we acquired additional land on the site to enable us to build ground car parking rather than the expensive underground car park the initial land size could accommodate. The landlord was now very cooperative, and we bought the additional land at the same rate we had paid a decade earlier.

We conducted the tender and the landlord was not the lowest bidder. However, we were not satisfied with the outcome and opted for post-tender price negotiations with lowest bidder. As a leader, if the return is greater than the investment, then it should be worth your time, so I met with the MD of the lowest bidder regarding the need to work together towards achieving the goal. I made it clear that we would require some target reduction if we were to proceed.

We assembled a joint team of engineers and contract staff to review all aspects and think outside the box. We opted for an open-book approach and by the end of the week the team had exceeded their target. Setting a target focused the team on the objective, as what gets measured gets done. We secured board approval and awarded the contract. A few weeks later, the Governor attended the foundation-laying ceremony and work began.[12]

A leader must ensure that the company gets into the best competitive position to maximize value from the company's spend. Extrication from a monopoly is good commercially, and I learned that to bring a monopolist to the table, viable alternatives and consequences (e.g. the landlord not participating in a tender) must be created. I could not accept NLNG being held 'hostage', irrespective of a 10-year arrangement.

NLNG Head Office Ground-breaking ceremony (2015)   *LNG completed Head Office (2019)*

Creating local job opportunities

In 2014, we decided to acquire six new LNG ships to replace our ageing fleet. We secured US$1.4 billion in loans for the acquisition,

---

[12] To my team members – Chima Isilebo, Bayo Adenrele and Ronald Okardi – my appreciation for their outstanding contributions towards enabling the building of the new head office.

one of the highest international financing deals arranged by a Nigerian firm at the time.

Korean shipbuilders (Samsung and Hyundai) won the tender at a competitive price. We saw it as an opportunity to boost local capacity and we entered into discussions with them to find win–win solutions before the final award.

One area of focus was skills building. In Nigeria, experience of building ships of this size was non-existent. We saw an opportunity to leverage contract to improve local knowledge in shipbuilding to global standards. Samsung and Hyundai agreed to train 600 Nigerians in Nigeria and Korea. The best 30 trainees joined Samsung and Hyundai in Korea to build the ships over a two-year period. They gained shipbuilding experience, and many have since become active in local shipbuilding (smaller ships) and in FPSO ship-fitting roles in Nigeria.

The second priority area was for local manufacturers to supply ship construction materials to Korea. The vision was to get local manufacturers to not only supply materials for this construction, but to be able to do the same for other future construction. Creating international linkage, market and foreign exchange earnings opportunities would grow local manufacturers and give them international credibility.

We commissioned a joint study to identify goods made in Nigeria that met or could be improved to meet global standards. It identified what local companies had to do to gain global accreditation. As a result, goods worth several tens of millions of dollars were exported from Nigeria to Korea for ship construction (cables, paints, anodes, furniture). Local companies established partnerships with Koreans for the first time and many of them also attained international quality standards.

The third area was for Samsung and Hyundai to provide simulators and build a training centre in Bonny. This would enable many more Nigerians to be trained in ship management. NLNG shipping staff and government port regulators are now being trained in foundation ship management training in Bonny, rather than in Europe.

The fourth area was to inspire building of a dry dock in Nigeria for ship maintenance. Historically, LNG ships are taken to Europe for maintenance, costing millions of dollars annually. The vision was that routine maintenance would be done in Nigeria with local labour, which would create employment. The spend would be retained in Nigeria and local competence would be developed.

We jointly appointed a Dutch company to review where a dry dock facility could best be located in Nigeria. From technical study, they identified Badagry, Lagos as most suitable, and identified a consortium of local and international partners[13] with the financial and technical capabilities to build such a dry dock facility on their own. We signed a tripartite agreement to underpin the vision.

With progress being made on the wider Badagry deep seaport, [14]the opportunity for the construction of the Badagry dry-dock project remains viable. A project of the magnitude of Badagry dry-dock (US$1.5 billion) requires painstaking work among the consortium and their technical partners, and takes many years.

I learned that delivering impactful value is not always obvious or guaranteed from initial outcomes. It requires having a clear vision and determination. One must think differently and believe that deeper change is possible. Sustainable impact occurs when the company's bottom line benefits, as well as society as a whole.[15]

---

[13] See   http://sifaxgroup.com/bsmec-emerges-lead-investor-for-1-5b-badagry-dry-port-project

[14] See www.shippingposition.com.ng/main-news/badagry-deepsea-port-project-in-progress-osinbajo

[15] To my team members – Temi Okesanjo, Victor Eromosele, Sam Orji, Ibrahim Lamah, Hambali Yusuf and Ekeinde Ohiwerei – my appreciation for their work with me to shape, nurture and implement the vision.

Chapter 5

Developing communities

What counts in life is not the mere fact that we have lived.
It is what difference we have made to the lives of others that
will determine the significance of the life we lead.
                              – Nelson Mandela (1918–2013)

Challenges of communities

Despite several decades of exploitation of oil and gas in the Niger Delta, the area remains hugely under-developed. Many towns and villages, especially in the swamp and mangrove parts of the Niger Delta, are without roads and bridges and can only be accessed by boat. Many do not have access to pipe-borne water or electricity. There are few industries and thus limited employment. As a result, the Indigenous people mainly undertake subsistent farming and fishing. The situation is compounded by recent widespread oil spills from the pipelines of oil companies (from its operation and from intruders/sabotage), which has made fishing and farming more challenging. As a result, poverty, huge unemployment and poor infrastructure are prevalent in Niger Delta communities.

The failure of government, companies and community leaders is at the heart of the under-development. It is a complex situation, and the major factors for the under-development are the lack of holistic master-planning and disciplined implementation by government.

Government failed to provide adequate basic infrastructure for the communities, despite hundreds of billions of dollars being allocated to the region's development since the 1970s.

Funds were corruptly frittered away by administrators[1] over the years, leading to poor execution of projects. In an October 2019 article, *The Economist* estimated US$582 billion had been stolen from Nigeria from 1960 to 2019.[2] The Nigeria Economic and Financial Commission estimated US $217billion was stolen from 1970 to 2008.[3] As a result, many projects have been abandoned, left incomplete or completed to poor standards. An example is Bonny–Bodo road, designed in 1970s by the federal government, but incomplete 40 years on. The road project has only now commenced after support by NLNG (see later in the chapter).

IOCs embark on community development as part of corporate social responsibility (CSR), and spend hundreds of millions of dollars on infrastructure in the villages where they operate. CSR normally complements government responsibility, but unfortunately there is little to complement in Nigeria. CSR impact thus seems feeble, especially due to the large size of the Niger Delta.

IOC community projects also suffer from corrupt activities by unscrupulous elements within the workforce, vendors and community leaders. While in Warri, I experienced community leaders' questionable commitment to the development of their villages. In the 1990s, as an SPDC buyer, a contract was signed to construct a school in one of the villages. To ensure value for money, the contract was split into two. One was for material supply (e.g. cement) and openly tendered for at competitive rates. The second was for labour; it was assigned to the community, to use community labour. The community nominated one of its chiefs to coordinate this.

A few weeks after the contract was awarded, the material supplier came to my office with the chief. He had delivered on site but was

---

[1] See www.premiumtimesng.com/news/headlines/358182-buhari-orders-forensic-audit-of-nddc-operations-since-2001.html

[2] See www.thecable.ng/chatham-house-582bn-stolen-from-nigeria-since-independence

[3] See www.today.ng/news/nigeria/efcc-217-billion-nigeria-illegally-38-years-187475

accosted by the chief, who asked him to halt further supply, but to bring his waybill so the chief could sign that all materials had been received, and the supplier could then submit this with his invoice to SPDC to get paid. The supplier was then expected to remit 80% of the payment to the chief. The remaining 20% was to be retained by the supplier, who would not need to supply any more materials. The supplier was concerned an SPDC audit could uncover this and indict his company. But as he was being threatened to comply or be attacked, he appealed to the chief to accompany him to SPDC to state formally what he had asked him to do. This was why they were in my office. I was shocked. I took them to my manager, who subsequently took them to the Divisional Director. At each turn, the chief insisted that his demand was with the knowledge of the King and other chiefs, and that they would all rather collect the money and share it than build a community school. The company leadership insisted on building the school and had to involve the state government to intervene; however, this led to a frosty relationship with the community elders over a long period.

Such insights reveal why, despite huge amounts allocated over the years by governments and multinationals, many projects in the Niger Delta exist only on paper or have remained uncompleted.[4]

One thing that I had learned from my father was that, rather than be despondent about the failure of government, individuals can make a difference. He committed his life to ensuring that his community developed and spent his time, money and influence to help build a school, electricity, pipe-borne water, a clinic, market halls, a town hall and youth projects in his village.

He physically participated in construction effort by loading building materials, laying bricks and lifting materials alongside hired construction workers. He could not stand delays to projects and would rather divert his personal income. During the teething years of the school, he contributed his own funds to paying the salaries of schoolteachers until government finally took over the responsibility.

---

[4] See www.vanguardngr.com/2014/06/nddcs-4000-uncompleted-projects

He also sought support for community projects at every opportunity. One was with Reuben, his sibling, who had secured employment and was entitled to a car loan. My father asked him to take the loan and give it to the school for payment of teachers' salaries. Reuben obliged and the school repaid the loan months later when its cash flow improved.

Over the years, I saw how his efforts, along with those of others, improved the community. With electricity, children were able to read late into the night; with pipe-borne water they no longer had to go to streams; with a clinic, mortality from childbirth was reduced; school enabled many children from the immediate and surrounding villages to get secondary education. In recognition of his efforts, in 1985 the King named him Akoluje of Okoro-Gbede. The honorary chieftaincy title meant 'one who leads the community to greener pastures'.

My father fell sick in 1994 and despite visiting many hospitals in Nigeria, the doctors were unable to diagnose his ailment. By 2001, when I was on overseas assignment, our family arranged for his travel to the United Kingdom for treatment. Within 24 hours at Aberdeen Royal Albert Infirmary, doctors carried out tests and determined that his kidneys had progressively deteriorated. They were upset that the kidney failure would have been preventable and controllable had the Nigerian hospitals correctly investigated and diagnosed the symptoms earlier.

Unfortunately, a kidney transplant – normally a standard procedure – was going to be risky due to his age and frailness. He commenced dialysis and after six months requested to return to Nigeria, where he endured further dialysis for eight months before succumbing to his illness. He passed away at his home in April 2003, two months shy of his 68th birthday.

I was sad and upset at his death, knowing that had Nigerian hospitals been better equipped and staffed, he would have lived longer. I spent months with him during his treatment in Aberdeen. He was proud of the progress of his five children and especially that we had absorbed many of his values, including integrity, hard work and excellence. He was a great role model and I remain proud of him.

He was interred in his hometown in a well-attended ceremony. The villagers sang *'Baba to feran wa o ti lo'*, which translates as 'The father who loved us has left us' and followed his casket for a long way. The town's Development Association recognized his integrity, non-discrimination and help to those in need. Students recognized his impact and legacy as an educationist.

My father loved his native land and it was obvious during his burial that the villagers loved him back. He had lived as a commoner but was honoured by his townsfolk in death. The fondness highlighted the impact he made on people and society. What struck me during his burial was that little was mentioned of his personal attainment (degree or career); the testimonials were about his impact on others and the community.

I learned that individuals can make a difference in society, and that service to others is a higher purpose than a focus on self. Success in life is not just about what one accomplishes, but more about how one can make others' lives better. One's achievements should be a stepping stone to helping others. I learned of the long-lasting impact one can have on the lives of others by sacrificing for the good of people and communities.

Making a difference in developing countries

In December 2011, a few days after being appointed CEO of NLNG, I visited the Bonny King at his palace. His Majesty Perekule XI (King Edward), had ascended the throne as Amanyanabo of Bonny in 1996.

Bonny was founded in the fourteenth century when the first settlers called it 'Okoloma' (curlew town), as the island was then full of curlews. In the fifteenth century, the Portuguese Atlantic slave trade growth led to Okoloma becoming the *entrepôt* of the slave coast. Later, the Dutch and then British took control of the slave trade in the region, and the British renamed the port 'Bonny'.

Bonny is a host community for NLNG's industrial and residential facilities. As I drove around, I was gutted by the glaring under-development and poverty I saw. This is where 10% of the world's LNG is produced and it generates 14% of the revenue of Nigeria. The island has hosted Shell and Exxon export terminals since the 1970s. This was not right.

The town reminded me of Warri when I lived there in the 1990s. These were towns that I had heard so much about while growing up. I had pictured oil cities like Houston, Calgary, Aberdeen, Kuala Lumpur or Muscat. I had expected to see high-rise buildings, industries, five-star hotels and shopping malls. I saw no such structures – just old houses with zinc roofs and petty traders on the road.

I began to think of how to work with the community, to generate a vision and framework for Bonny's transformation. I visited historical projects that were benefiting the community, such as electrification, which enabled Bonny to be one of the only Nigeria towns to enjoy 24-hour electricity. Water and road projects were being built by a Joint Industry Committee (JIC) between NLNG, Shell (SPDC) and ExxonMobil. I also visited the NLNG projects, including the nature park, bakery, residential complex, health centre and vocational college.

I also visited project sites that were uncompleted or not yet started, including a housing estate, school and sports complex. I viewed these as unfulfilled promises. Based on my experience in the Niger Delta, I knew such say–do gaps would undermine trust. Without addressing past commitments, there would be little chance of cooperating on any new vision. I got my team focused on completing these projects.

While the completed historical projects were a great infrastructural foundation, I reflected that there was a need to diversify and build a thriving economy. Sole dependence on few IOCs was not sustainable for such a fast-growing town of 215,000 people (2006 census). We had to enable economic development and the job-creation capacity to meet the town's socioeconomic needs. It needed industries, agriculture, recreation, tourism, housing and more.

With these thoughts in mind, I began to discuss with King Edward the concept of a long-term vision for Bonny. I suggested that that it would be a legacy for which future Bonny generations would remember him. I described the global influence now wielded by the rulers of Dubai and Singapore. The King was positively disposed – it was a meeting of minds.

It is key for a community to take the lead and own the vision for its development. We discussed setting up a Community Foundation as the vehicle to execute the plan, rather than NLNG. It was to ensure the community owned the projects and not feel they were NLNG projects. This had made past IOC projects unsustainable.

The King was apprehensive if it meant NLNG would be walking away. I assured him this was not the case, as we would be on the board of the Foundation and would provide financial contributions.

To enable the registration of the Foundation and commence community engagement, we hired a reputable firm, Accenture, which arranged sessions with various groups (women, youth, elders). At each session, the King articulated the vision we had discussed.

We translated the vision into a 25-year Bonny Master Plan, with four pillars: (1) human capital; (2) economic diversification; (3) infrastructure development; (4) environmental sustainability. The expectation is that by 2040 Bonny will look back with pride at the impact of the master plan on its development and wellbeing.

The NLNG board was concerned about whether the community could handle such a level of development, but finally approved support for the master plan. N3 billion annually over 25 years (equivalent to US$210 million) was approved towards the execution. I also secured SPDC's support of N600 million annually (equivalent to US$42 million).[5]

The expectation is that additional financial contributions towards the master plan will come from contractors who would be working on the Island, such as the major engineering and

---

[5] See   www.orientenergyreview.com/oil-and-gas/under-new-mou-shell-nlng-provide-n3bn-for-bonny-development

construction contractors who would work on expansion projects such as Train 7. It is also expected that when progress is made, development bodies (the United Nations Development Program, Africa Development Bank, Department for International Development) and the government would be inspired to support the execution of the master plan with additional finance.[6]

The Bonny Kingdom Development Foundation (BKDF) was registered[7] as a vehicle to execute the plan. However, in 2017 some community members instituted a court case, challenging the nominees the King had named for the BKDF. The legal action delayed the commencement of BKDF and the execution of the master plan. It is another example of how people in Niger Delta put personal interests above those of society, creating obstacles that stifle development.

To support the execution of the master plan, I secured board approval to provide N60billion (US$170 million) towards construction of the uncompleted 38-kilometre Bodo–Bonny Road to link Bonny Island to Port Harcourt on the mainland. Absence of road access into Bonny meant that transportation of people and goods was limited to waterways, adding significant costs and with security (piracy and militants) and safety (poor quality boats) implications for inhabitants. The support enabled the construction of the road project.[8]

I also secured the board's endorsement for change of strategy for ownership and management of infrastructure built in the community by NLNG over the past 20 years. This included the vocational college, electrification facilities and water infrastructure. These had been managed by NLNG since commissioning and the community did not feel any sense of ownership or accountability.

---

[6] The video of this is available at www.youtube.com/watch?v= Rmf0-zUJAUo and the details of the Foundations at www. bonnykingdomfoundation.org/our-work-2/master-plan

[7] See www.bonnykingdomfoundation.org/about/governance

[8] See www.julius-berger.com/references/bodo-bonny-road-rivers-state

This was unsustainable, particularly based on the experiences of other oil and gas companies in the region over the years.

Community ownership would bring about a change in how the community approached these facilities, leading them to properly utilize and protect the facilities. In addition, managing such infrastructure would help the community (through the Bonny Foundation) to develop managerial capabilities.

Such a level of planned development was unprecedented. It was the largest single corporate social responsibility undertaken by any company in Nigeria. It served as an example of how individuals can make a difference by organizing communities to become active in their own development and to be less reliant on a few private companies. This is one model for the development of the Niger Delta.

Making an impact in areas of influence is important for career success. A leader must create a vision and pursue implementation. It requires thinking outside the box about improvements that have an impact beyond a company. While delivering the bottom line and maximizing returns are expectations, leaders also need to seek opportunities to positively impact society. A firm that is not relevant to the society in which it operates is not truly sustainable.[9]

Engaging and mobilizing the support of enlightened key stakeholders (such as the traditional leadership) is critical for community transformation. Enabling community groups (e.g. women, youth) to play an active part in the long-term development of communities is crucial for ownership. Using experts (consultants) outside an organization's area of expertise is vital to enable success and provide independence.

---

[9] To my team members – Ifeanyi Mbanefo, Fola Olanubi and Segun Sowande of KPMG – my gratitude for their tireless work with me to achieve results in the community development.

Chapter 6

Vision

The best preparation for tomorrow is doing your best today.
– Harriett Jackson Brown Jr (b. 1941)

Keeping the team on track

I walk into the meeting room in Bonny and see that the team is downcast. They are aware of the setback at the board meeting yesterday. The organization changes on which they had worked so hard had not been endorsed. I thank them for their effort and encourage them not to be demoralized. I assure them that we will ultimately achieve our objective.

One of them asks, 'Would you disband the team?' I tell them a story of a young marketing executive who one day hands over his resignation to the CEO. An idea that he had pushed, which cost $10 million, had failed. He was contrite and accepted responsibility for the loss. The CEO rips the letter up and says, 'Your resignation is not accepted. We have just spent $10 million to train you, so why would I throw away that experience?' I look at each of them and say firmly, 'No, I will not disband the team. There are lessons to be learned, but we are yet to achieve our goal. You still have work to do, so get on with it.' Their eyes brighten and the mood in the room changes.

Failure challenges commitment. To achieve a goal, one may have to go through failure, but it is vital to be able to pick oneself up and continue. Thomas Edison tested 6000 materials as possible contenders before inventing the light bulb, and famously said, 'Genius is 1% inspiration and 99% perspiration.'[1]

---

[1] See http://edisonmuseum.org/content3399.html

A leader must motivate their team through the disappointments that may arise during change, and through the onslaughts along the way. The leader must renew the team's commitment and keep the team and stakeholder focused on the vision and the goal, and on track to achieve it.

Being ahead of the curve

A leader must always be ahead of the curve to anticipate what is to come and prepare the organization for it. They need to scan the external and internal environment, to look at patterns and connect the dots. Anyone who leads an organization to stumble into a future reality will have to fight fires to respond to the new reality, which does not give confidence to stakeholders.

By 2009, I was one of the most senior Nigerians working for Shell. Company leadership in The Hague reviewed performance and potential and identified me to become MD/CEO of NLNG on secondment from my position as a Shell VP in Sub-Saharan Africa. I proceeded on a one-year onboarding programme, which included 'shadowing' CEO/MDs of some LNG plants and some training. I also spent time in the head office meeting with key stakeholders.

My appointment to the Nigeria LNG and Bonny Gas Transport Limited boards in December 2011 was a proud moment. Looking back at the start of my career in a storeroom, I could not have foreseen these boardroom appointments. I had joined SPDC in a non-core function, and unlike the Petroleum and Engineering departments, which usually produced MD/CEOs, none had emerged from a non-core department since I had joined SPDC.

NLNG produces liquefied natural gas (LNG), condensates and LPG. LNG is a safe and clean energy source. It is gas, cooled to $-160°C$ and pressurized to turn it into liquid so that large volumes of it can be shipped to distant places, as liquid occupies $1/600^{th}$ of the space of gas. When LNG reaches the location of customers, it is converted back to gas by a reverse process and used to generate electricity, heat homes and fire industrial engines. It is also fast replacing gasoline and diesel throughout the transportation sector.

BGT is a shipping subsidiary and owns 20 LNG ships, one of the largest fleets in the world.

NLNG is the most successful indigenous company in Nigeria. Its six production lines (trains) produce 23 million tonnes per annum (mtpa) thus supplying 10% of global demand, initially to customers in the United States and Europe. It was the fastest built LNG plant and the fourth largest in the world. It was listed in 2014 by Nigeria's Federal Trade Ministry as the top-ranked home-grown company, having accounted for 7% of the country's GDP.

The NLNG model has been a success in Nigeria – more than the upstream JVs and refineries. There are a few reasons for this. One is the shareholding structure, where the three IOCs own 51% while the NOC owns 49% (unlike joint ventures, where NOC owns 55–60% and refineries are owned 100% by NOC); this ensures that decision-making balances international standards and local expectations. In addition, the independent board ensures quick decision-making on investments, operations and contracts, devoid of bureaucratic delay. The company is also able to fund its operation and projects from its own funds or external borrowing and is not subjected to the vagaries of government delays in budgeting and cash call challenges. It is also important that its managing directors have mainly been from an IOC – not because IOC staff are better than NOC staff, but because IOC staff do not owe their future to government officials and politicians (unlike NOC staff), so are able to more easily take decisions on merit and resist political pressure.

On my appointment to the NLNG Board (2011)

As NLNG was already a successful company when I resumed work there, I set out three main objectives, as it is important for a leader to focus on a few priorities: (1) sustain historical performance; (2) take the company to the next level; (3) make the company an inspiration for the country.[2]

By the end of 2012, my first year as CEO, we had grown revenue above US$11 billion (from a historical average of US$6 billion per year). We achieved this record revenue by producing higher volumes (10% more than the historical average). We sold LNG at higher price in new markets that we had secured in Asia, rather than for lower prices in the United States and Europe. We also benefited from the generally higher energy prices in the market.

We declared the highest dividend in the company's 13-year history ($5.6 billion) to shareholders. We paid a generous bonus to the employees. Shareholders, directors and staff were pleased.

Yet while stakeholders were celebrating, I was worried. I had analysed the company's internal and external factors, and many indices were not looking good for the company's future.

First, global economic indices showed a global debt crisis and looming slowdown of the global economy, including our niche (most profitable) Asian market. Demand for our products would likely decline.

Second, new LNG plants were emerging – Qatar (77 mtpa) and Australia (80 mtpa) – and the United States was now able to competitively extract shale gas, which would double global reserves. This meant competition was increasing and customers would have more options.

Third, alternatives (e.g. renewables) were growing, with lower costs. In addition, a global climate change drive was striving to reduce the use of coal, oil and gas, due to the by-products they produce: CO_2 (carbon dioxide) and CH_4 (methane). While

---

[2] See www.linkedin.com/pulse/what-did-babs-omotowa-do-nigeria-lng-taiwo-obe

renewables in the energy mix were relatively small (<5%), technology was quickening its growth. Even NLNG shareholders were investing in renewables.

Fourth, the supply of our raw material (gas) was threatened due to growing demands to use gas for domestic electricity generation, which was ranked higher by the government and citizens. As a result, we would be facing challenges obtaining raw gas in the future.

Fifth, we were facing a more complex business environment with the imminent end of our pioneering incentive tax holiday and emerging new regulations (e.g. the Petroleum Industry Bill).

Finally, oil price analysis highlighted that the prevailing high oil price ($140 per barrel then) was likely a peak. Oil prices had cycled since 1970 and the lesson was that such a peak would likely be followed by a crash. As LNG prices are linked to oil, the interpretation was that our LNG prices were about to fall.

From this analysis, I saw headwinds. This contrasted with the optimism of stakeholders. It is a dilemma when a leader sees a different future from the rest of team. It was important to introduce stakeholders to this challenging reality and the need for speed, as the analysis indicated we had to act fast. In such a situation, one needs to create a *burning platform* in terms of a strong case for change and the need to act.

It is important in any change process to ensure that the issues are well understood before strategies are developed. I began to share my analysis with the board, management, employees and at external conferences. I warned about a 'closing window of opportunity' and the 'eroding market share', and the consequent need for the company and government to act fast against the imminent threat.

Some understood the challenge while others believed the reality would remain. We were like a football team at the top of the league, whose over-optimistic supporters were blind to burgeoning problems such as a lack of youth development, unhappy players or growing debt.

Addressing the headwind

I began to think of the plan to address the headwinds. During this period, I often found myself waking up in the middle of the night, thinking of the challenges that the company was facing, and how to bring others on board to address the issues. I had a notebook and pen beside my bed, and as thoughts and ideas came to me during the nights, I wrote them down. I wanted to be sure I did not forget any thoughts when I woke up the next morning. I get some of my creative thoughts during such periods.

In change, it is important to create narratives that can inspire stakeholders. I developed an 'A to B' pathway – 'A' being the current situation and analysis of key metrics (reliability, utilization, costs) and 'B' being the desired vision, where we want to be in five years. The end goal must be clear, different and inspiring. Graphically showing the 'A to B' picture became a powerful way to communicate and provide meaning for stakeholders on what we needed to achieve and the path to get there.

To dig into the analysis, engage stakeholders, develop and implement solutions, I set up four teams under an overall Transformation Improvement Team (Project TIM). For each team, I selected an external consultant to provide expertise and appointed a management team member to steer and guide.

Organization culture

The management team, composed of managers seconded from shareholder companies, had influenced the culture in NLNG over the years. This led to a situation where the company had multiple cultures, with each department having a culture akin to that where the seconded manager had come from. For example, Human Resources was led by NOC seconded staff, and as a result the culture there was akin to civil service, where workplace harmony was important. Production was led by IOC seconded staff, and run with international standards, with performance important.

My working experience had been in companies where staff were aligned to the same objectives and culture. In those companies, there were common purpose, common management systems and common ways of working. I was shocked by the absence of integration at NLNG.

From my observation during my first year and staff surveys, it became clear that without tackling culture, we stood no chance against the impending headwind. Examples were silos between units; the command and control style of leaders; waste tolerance; a poor work ethic, a rumour-mill environment; failing communication; and staff not speaking up. These were worsened by seconded staff not working well together, as they had more allegiance to their parent company.

Some staff questioned the need for change, asking 'How come our historical performance has been decent despite the culture?' I explained that in the first 10 years, NLNG was in a project phase (construction), which meant the huge capital budget associated with construction and the increasing production volume capacity masked inefficiencies. The company was now entering a new phase – asset management – where the operating budget would be predominant, and production volume capacity stable. Operating budget are much smaller than capital budgets. As a result, the inefficiencies of the prevailing culture would be magnified, if nothing changed. Past successes are no predictor of future performance and can trap one in a rut.

The stifling culture would limit NLNG's performance and ability to confront imminent challenges. Management guru Peter Drucker stated that 'Culture eats strategy for breakfast'. Louis Gerstner, former IBM CEO, stated, 'culture is everything'. With this as background, I took it upon myself to personally lead the efforts on the culture workstream.

We had to develop a culture in tune with global best practice. We engaged 1300 staff in 50 sessions and used anonymous online surveys to discuss the prevailing culture and come up with ideas about what the new culture should be. Ten new behaviours emerged as 'NLNG culture', including collaboration, listening, empowerment, fact based, communication, continuous improvement, urgency and role model.

Due to the involvement of staff in generating new culture, their buy-in was high. They generated innovative ways to implement it, including a 'culture moment' at start of meetings, where a staff member would describe a culture element and share how they had demonstrated it, or the challenges faced. A 'feedback' culture was introduced for staff to provide feedback on how well bosses were demonstrating the new culture and how they could inspire team members. On communication, 'townhall' briefings were implemented to enable staff to ask questions. As not all staff were comfortable speaking in large groups, I held lunch sessions, in groups of seven to ten, to enable them to interact directly with the leadership and share ideas in a private setting.

Changes in the behaviour of leaders had the biggest impact, as staff tended to emulate them. The focus included changing their attitude from 'master/servant' to 'inspiring/coaching', from 'telling' to 'listening' and from 'silos/fiefdom' to 'collaboration'. One of the biggest concerns was the commitment of leaders. Staff observed that some leaders who had benefited from the past culture were exhibiting passive resistance. These leaders acted differently in public, compared with the way they behaved privately. There were also concerns that some staff were using the new behaviour elements to game the system, such as reporting that their leaders were not listening when the staff were being held accountable for performance issues.

During this period, a secondee manager once stormed into my office, visibly upset. He said, 'These boys are disrespectful, under a culture of speaking up.' 'What happened?' I asked. He said, 'In a meeting that I chaired, one of the staff had the temerity to pick holes in a solution I proposed, and went to suggest something different. This is insubordination.' I empathized and replied, 'I will ask the team to share with staff some techniques to bring out ideas respectfully.' However, I also shared with him the story of a company where some young staff picked up the idea of the xPhone from a supplier. At a later meeting, where the marketing leader proposed a different idea, they suggested the xPhone, based on the company's culture of speaking up. After review, the company went with the choice and market capitalization quadrupled. I asked the manager to reflect on this but sensed that the disrespect aspect continued to play in his mind. I encouraged him to think of how to achieve both the objective of speaking up and that of being respectful and to share with the team. He looked flustered when he left my office, but during his send-off years later, he described finding the discussion useful. He had later changed his approach to staff speaking up and found their contributions valuable. He committed to taking the approach back to his parent company.

With unrelenting change management efforts over three years, there was evidence of change. Staff began to speak up freely, as evidenced by feedback, silos were eliminated through structuring and integrated work processes, appointments were made through advertising vacancies and panel selection occurred on merit.

As culture change takes time, building and maintaining trust across the organization is critical. Culture is the foundation for an organization's success and involving the staff through the culture-change process is key to their commitment. Ensuring supervisors are onboard with the change is very important, as they are usually beneficiaries of existing culture, and the examples they set regarding the new culture have a major impact on staff.

Commercial agility

The commercial workstream focused on finding new high-value LNG markets in Asia (Japan, Korea, China). Historically, 40% of our LNG was exported to the United States. However, prices there had fallen by 65% with the growth of shale gas following technological breakthrough in its extraction. Customers there now had access to cheap domestic gas and no longer needed imports. The United States was changing from a gas-importing country to an exporting one due to improved ability to extract local shale gas cheaply. It was looking to build LNG plants, and to export to Asia to take advantage of prices there too. A customer was now becoming a competitor. As a result, we aggressively went after the Asia market and secured some favourable deals.

Our other product was liquefied petroleum gas (LPG – propane, butane, locally called cooking gas). It is produced as a by-product during LNG production and is supplied to residential homes in cylinders. The cylinders are connected to gas ovens and cookers through hoses. When the cooker knobs are opened, the gas flows through the hose into the cooker, which is lit and used to cook. We were producing 800,000 metric tonnes (MT) of LPG annually and supplying only 100,000 MT to the domestic market. The rest was exported.

An NLNG survey of the market had shown Nigeria was using only 10% of LPG potential. Locals were using alternatives (wood, kerosene), which cost more than LPG and negatively impacted the environment (deforestation with trees being cut as firewood) and health (people inhaling smoke from burning wood). Kerosene negatively impacted government finances, as the government was subsidizing it to the tune of tens of millions of dollars.

We could continue to sell LPG internationally, but there were advantages to helping Nigeria develop its LPG market. We would add value to the country and enhance our reputation. The government would benefit (ending subsidies) and both citizens' health and the environment would improve (reducing deforestation and CO_2 release into the atmosphere).

Our consultant's (KPMG) review of other experiences highlighted that Indonesia's programme had led to 50 million users of kerosene switching to LPG. It included free distribution of cylinders and awareness. The subsidy on kerosene in Indonesia before 2007 was 10–18% of the national budget. By 2011, it had fallen from US$12 billion in 2001 to US$3 billion.

The review highlighted that with the right actions, the Nigerian market could absorb a million metric tonnes. KPMG identified the bottleneck as poor infrastructure to transport, store and distribute LPG. Expansion would require developing supply chain capacities and issuing free cylinders to people. Raising citizen awareness would be a key focus to ensure they understand that firewood and kerosene were more expensive than LPG, as well as the damage to their health posed by firewood. This was important, as a similar scheme of distributing free cylinders in Nigeria in the 1990s saw many rural people selling the free cylinders for money, due to their poverty level.

In the short term, I obtained board approval for NLNG to increase LPG volumes to the domestic market from 100,000 to 250,000 MT. However, as the challenge was across the supply chain (supply, storage, distribution, retail, cylinder), it would require more than NLNG. We needed to enlist government. I initially pitched the plan at ministerial level (Environment, Petroleum). The ministers agreed, but a turf war between the ministries, rather than collaboration, prevailed. We needed higher sponsorship. In Indonesia, it was led by the Vice President. In 2015, with a new government in Nigeria that was better coordinated, I approached the Vice President, Professor Osinbajo, to champion it. He agreed and assigned relevant aides.

As Chairman of the National Economic Council, he had the mandate to bring ministries together. At his behest, we convened a stakeholder meeting with the KPMG report as the basis. He appointed a Special Adviser on LPG to coordinate the implementation of the solutions. Progress began to be made, including a new policy on LPG cylinder ownership, resolution of VAT anomaly (locally produced LPG attracted 5% VAT whereas LPG imported from other countries was VAT free) and distribution of free cylinders.[3]

Our efforts and the enrolment of stakeholders, including senior government officials, enabled renewed concerted efforts towards making sustainable changes in an area with health, environmental and financial benefits for the nation.

Organizational structure

Benchmarking NLNG against other LNG plants globally and IOCs in Nigeria highlighted inefficiencies. There were two organizations

---

[3] See https://africachinapresscentre.org/2019/06/10/lpg-can-create-2m-jobs-in-nigeria-says-vp-osinbajo

existing within the company, with duplicate functional teams: one in the Bonny plant and the other at the head office in Port Harcourt.

The need to remove duplication and integrate them was glaring. Organizational design on the basis of assets (production, pipeline) was selected due to the asset management phase the company was in. This was straightforward but became the most emotive of the workstreams. Such company-wide redesign had never occurred before as NLNG's first 10 years had been focused on constructing the 6-Trains. Staff who were beneficiaries of the existing setup (e.g. favoured in appointments) did not welcome any changes that would disrupt the status quo.

I knew the critical path involved enrolling shareholders. I directed an engagement strategy where the team provided design details, but engagement was to be led by the most senior seconded staff from shareholders (e.g. for NOC, their most senior secondee, NLNG's Deputy Managing Director, led the team and similarly for the IOCs). The team received comments and the general feedback was of support, but also some resistance.

The reasons for resistance were not too far-fetched. First, proposals were underpinned by analysis and logic, but hadn't reflected the historical antecedents, which had limited the number of NLNG staff in management positions and had ingrained NOC and the IOC secondees in the company set-up as 'checks and balances'. The proposals had provided for additional management positions for NLNG staff, which would also provide headroom for the company staff.

Second, until 1994, NLNG Managing Directors (MDs) were NOC secondees. However, just before taking the final investment decision (FID) to start construction, the IOCs insisted that this be changed before they would support it. This was agreed to by the Head of State and subsequently MDs were secondees from Shell. As a result, the NOC viewed any proposal on organizational restructuring with extra scrutiny.

Third, some NLNG staff were apprehensive the restructuring may lead to loss of privileges, as they could be redeployed from their current roles to less attractive roles. As a result, they created deliberate falsehoods within and outside the company, including a narrative that there would be forceful staff layoffs, whereas the plan was in fact for voluntary redundancy. Based on informal nexus that existed between some company staff and NOC staff, they created erroneous perception that the changes would provide more opportunities for IOCs than for the NOC. All these manoeuvres eroded trust in the reorganization efforts.

I was not initially aware of these issues, to address them and allay the fears of concerned shareholders, or if necessary to make changes to the proposals. As a result, at a board meeting in December 2012, the proposal was not endorsed. Whilst most directors were supportive, the chairman stated that since the board tradition was to take decisions by unanimous consensus, the lack of support of a shareholder meant we had to halt it. I was upset but managed to contain my disappointment during the meeting.

This was a major setback and the reason for my meeting with the team that morning in Bonny. I announced to staff that we were not going forward with the restructuring as originally planned. I did, however, make it clear that it would not be business as usual. Changes were needed to strengthen the company's ability to be competitive in the future.

I was determined, and was not going to be turned around. I had learned from experience at the farm not to back down due to setbacks. If we failed to achieve a good harvest in one year, we went back the next year with the same crop but with improvements in our methods. I met with my management team to review what had gone wrong and revise our strategy, using the knowledge we had gathered. When in a rabbit hole, it is important not to dig deeper. We decided that instead of a wholesale change approach, we would address contentious issues and represent our solutions to the board in phases.

At the next board meeting in 2013, I presented the first phase and obtained approval. The case for change was clear and the proposals in that phase were not in contention. I presented other phases at future board meetings and gradually obtained support. By the end of 2013, over 95% of original proposal had been approved, leaving few contentious elements. The approved changes were adequate to achieve our goal. When one meets obstacles on a journey, like a mountain, one can bore through, dig under, climb over or go around.

One action I took as part of the restructuring was the redeployment of 80% of managers and department line heads. Many had occupied the same position for a decade and become set in their ways and comfort zone. Staff reporting to them had become subservient and owed more of their allegiance to the managers than to the company. Most staff were disillusioned with the 'sit tight' syndrome, which also stifled the growth of younger talented staff and created a glass ceiling.

The scale of redeployment was unprecedented and led to a lot of positive change in the company, especially inspiring staff and ending the myth of 'sacred cows'. But some viewed the changes as too sweeping, as there were few areas where square pegs ended up in round holes and struggled. However, on balance it was the right thing to do and a leader must be comfortable to make calls even when the picture is not complete and the decision will not be 100% perfect. I later learnt that the speed and scale led staff to call me 'Hurricane Babs' behind my back.

The lesson is that there are usually historical reasons for an existing structure in a complex organization. A new leader trying to make changes is best served by fully understanding that context, and factoring it into the change-management and communication strategy. It is important for new leaders to deliberately go out of their way to seek out what they may not be hearing or is not being said.

Organizational change is never easy as it affects people and emotions. There is no perfect approach or structure, so it requires sensitivity to stakeholders, addressing any contentious issues and extensive communication. It requires re-strategizing when necessary. A leader must ensure that the goal remains the focus despite any setbacks, and that despite disappointments, confidence should be shown to staff through the process to motivate them.

Cost optimization

For the first decade since its inception, NLNG was in 'project' mode, constructing LNG trains. The period was characterized by huge *capital* expenditure spend. Less attention was paid to the relatively lower *operating* expenditure spend, as the plant construction was of higher priority.

By 2011, when I resumed as CEO, construction had been paused and we were in an operational phase. We benchmarked our performance, which showed that NLNG was not in the top quartile. In several areas, performance was lower yet we were incurring more costs than others. We had to address these issues to be competitive. Cost reductions were needed as we were going into a period of lower revenue, based on analysis.

We instituted a cadence approach for opportunities across the business. Cadence is a disciplined improvement methodology, centred on listening and understanding opportunities, followed by agreement and actions, stitched together with effective communication. Opportunities identified included the laying up and selling of vessels and extending major shutdown maintenance to five-yearly rather than four-yearly. Implementing several opportunities identified led to significant reduction in operating costs.[4]

One challenge was costs that directly impacted staff. I led by reducing my own entitlements. Discontinuing the practice of all staff attending an overseas course every year was emotional. We changed to local training and when not available (e.g. a technical area), relevant staff could attend training overseas on rotation. This was consistent with global practice.

During this period, a staff asked at town hall meeting in 2015, 'Are we getting additional compensation this year?' I sensed he was not alone, as the hall fell silent. I was visibly disappointed and responded, 'What aspect is not clear about the headwinds? Our revenue has dropped 50% and the future is bleak. We need to

---

[4] See www.nigerialng.com/media/Publications/2019_FACTS_AND_FIGURES_ON_NLNG.pdf

reduce costs to cope with revenue drop. This is not a time for salary increases.' After the session, my deputy came to my office and said that if I could have seen my face when I responded, my disdain was palpable. Staff feedback was that although the question may have been insensitive, I should see it as part of the speaking-up culture that we were encouraging.

Between 2012 and 2014, staff struggled to come to terms with why, in a period of revenue growth, we needed to reorganize and reposition the company. However, by 2014 oil prices had fallen by 80%, from $140 to $30 and so did LNG prices. By the end of 2015, NLNG's revenue dropped by 50% to US$6 billion and fell even further in 2016 to less than US$5 billion compared with US$11 billion in 2012.

Anticipation of the headwinds in 2012 and confronting them with bold measures, including reducing costs by 40%, enabled NLNG to remain profitable during the period of low oil prices from 2015. It ensured delivery of decent shareholder returns despite reduced income. In contrast to some other oil companies, which resorted to staff layoffs from 2015 as a knee-jerk response to the lower oil price, NLNG did not lay off any staff. This was due to the foresight and vision that had operated since 2012.[5]

Staff and shareholders later came to appreciate this foresight, which helped to manage the years of lower revenue. During the company's 30th anniversary celebration in October 2019, it was a major talking point.

A leader must focus on where value will be for the company. They must look ahead to ascertain the picture of the future, even if it is daunting or different from what others see. The leader must articulate the picture and obtain buy-in from stakeholders.

Leadership is not about being popular, but about doing the right thing for the greater good. A good leader is not one who

---

[5] To my team members – Aka Nwokedi, Kudo Eresia Eke, Patrick Olinma, Temi Okesanjo, Chima Isilebo, Victor Eromosele, Solomon Folaranmi, Fola Olanubi and Abdul Ahmed, and the external Schlumberger Consulting Partha Ghosh and Tosin Iyi-Ojo – my appreciation for their unrelenting efforts and the results achieved.

reaches a decision by taking a poll of majority opinion, but rather a moulder of consensus who takes the position not because it is expedient, political or popular, but because it is the right choice.

An example of what I learnt from doing the right thing occurred in 1986. A Mr Dimke, from Eastern Nigeria (Ibo ethnic group), who worked for Tate & Lyle Sugar, was transferred to Ilorin. He had obtained a recommendation letter to enable his wife, a teacher, to transfer her services from East to Kwara State.

The Executive Secretary of School's Board refused to honour the letter, on the basis that Dimke's wife was from a different ethnic group (Ibo) and that indigenes of Kwara State (Yoruba ethnic group) were yet to be employed. Nigeria comprises three main ethnic groups (Hausa, Ibo, Yoruba) and ethnic sentiments and favouritism still linger.

After several months, someone asked Dimke if he had taken the issue to my father, who was Chairman of the School Board. My father was disappointed to learn of the ethnic discrimination. He explained he had experienced similar attitudes when he was in Northern Nigeria and told Dimke to leave it with him. Dimke had worked hard to become a Manager at Tate & Lyle in Ilorin, and my father believed this should not be overlooked.

Two days later, Dimke was in his office in Ilorin when he was informed that he had a visitor. He was shocked that my father had left his high-ranking office to come down to see him. As soon as they greeted one another, my father gave Dimke the acceptance letter for his wife. Dimke was speechless.

Chapter 7

Integrity

The world will not be destroyed by those who do evil, but by those who watch them without doing anything.
– Albert Einstein (1879–1955)

When no one is looking

I have just returned to my hotel room after delivering a speech at a conference at Eko Hotel in Lagos in December 2012. A marine contractor CEO, an expatriate, knocks and walk in. I assumed when he made the appointment that there were issues with the boat services the contractor provides to NLNG. I offer him a seat and he starts to talk about the challenging business environment in Nigeria such as uncertainties with fiscal and monetary policies and unfriendly regulatory agencies. I empathize and mention that we are facing the same issues and working on how to operate more effectively.

After several minutes, he starts to take his leave. He hands me what he says is a Christmas present. It is an envelope. I check and see wads of dollar notes. I am in shock: I did not expect an expatriate to do this. I hand it back and tell him that 'cash gifts are inappropriate' to be offered by a contractor to company staff. He seems surprised. 'Have you read NLNG Anti-Bribery and Corruption (ABC) policy?' I ask. 'If your company is making so much profit that you can be offering cash gifts, you should instead offer NLNG a reduced rate for the marine vessels you are providing to us.'

He leaves the room shamefaced. I open my laptop and record the details of what transpired into the online ABC reporting tool. I contact the Shell Director and we agree to investigate. The confidence and audacity of this approach worries me. It seems it is something he is used to doing. During the internal investigation that follows, he denies he offered me cash.

Without establishing malpractice through internal investigation, I was unable to take further action, such as sanctioning the contractor or reporting the matter to the police for investigation and for possible prosecution. I later reflected that perhaps I could have approached this differently. For instance, should I have collected the envelope as 'evidence' or found a way of placing the conversation on record? Should I have insisted on meeting him only in a public place in full view of others, or had someone else with me? CEOs regularly have to meet stakeholders one on one, and sometimes at short notice. The key thing is to always act in a transparent and ethical manner, even when no one else is present, and to avoid the perception of compromise.

However, this was the first and last time anyone attempted to bribe me in the role. Perhaps he shared his experience with other contractors.

Back in the office, I instructed the Procurement Manager to send a copy of our ABC policy to all our contractors and to request them to acknowledge receipt and share with their staff. I also asked him to renegotiate the day rates of all vessels on contract. We achieved double digit reductions in rates and saved millions of dollars.

Employees list honesty and integrity as the top attributes they seek in a leader. Integrity is consistency of character, even when no one is watching. It is about doing the right thing, being upright, truthful and having moral principles. It is an uncompromising adherence to ethical principles and values.

My ethical values were shaped by Judeo-Christian teachings, which I learned from the church and my parents while growing up. We learned about the divine hand of God in our lives. I learned about honesty from the biblical ten commandments and learned it in practice from observing my father.

His contemporaries recognized him as frugal, honest and a stickler for due process. Some relatives and friends were not pleased that he would not bend rules for their employment or scholarship. Some felt he was rigid and became irritated with his inflexibility. He believed everyone deserved an equal chance, and that with hard work anyone could achieve their dreams without having to abuse due process or for him to abuse his influential position.

I observed during this period that tension arises from doing the right thing, and I learned that people could view it as not wanting to help. I learned that one must do the right thing always and irrespective of how small one's sphere of influence may be, as that also gives one a moral right to demand the same from those in leadership positions in society. It is hypocrisy to be dishonest when no one is watching, but to publicly espouse the need for honesty.

In 1988, my brother Dele transferred from the Education Ministry to the University of Ilorin. Due to an administrative lapse, there was a delay in implementing his pay stoppage. He continued to receive his salary three months after his disengagement. When a Ministry staff member brought this to my father's attention, he was furious and demanded that Dele repay the money the next morning. As Dele did not have the funds, my father wrote him a cheque for the repayment. He asked him to write a letter to the Ministry to acknowledge his misdoing and request a kind review. He made sure Dele fully repaid the loan.

I watched my father's adherence to integrity, honesty and fairness in both his public and private life. These values, which I learned from him, became traits for which I also became known in my career.

Beating the odds early on

I had a challenging experience in 1995 when I stood in for my supervisor during his leave. His duties included the transport operations along with the warehouse and yards, which were my primary responsibility. I reported to the acting manager.

One day the acting manager instructed me to issue materials worth hundreds of thousands of dollars to a contractor. I asked for an authorization document. There was none but he wanted me to issue the materials anyhow, based on his verbal instruction. I returned to my office to check whether there was any company procedure that provided for such verbal instruction from a manager without documentation; I found none. The next day, I returned to his office to inform him of this.

I pleaded with him to sign authorizing me to act. He refused and asked whether I was being insubordinate. He stated that I should simply obey his instructions since he had authorized me. If I refused to act on his order, I would face the consequences. He threatened to stall my progress in the company.

I requested that he wait for my supervisor to return from his leave, or to have someone else act for my supervisor. I mentioned that I had only been in the company for a few years, and maybe others more experienced than I would be in a better position to carry out his instructions. He sent me out of his office. I returned to my desk and continued working. I avoided the acting manager until my supervisor resumed from his leave two weeks later.

True to his word, he ensured I suffered consequences. I was dropped in performance ranking and estimated future potential in that year's appraisal. This was not the outcome I believed I deserved based on my output. I was, however, not perturbed as I knew there was not much that I could do about it and so continued to focus on my work. In subsequent years, I regained my relative performance and potential.

Another challenging experience occurred in 1996. I worked as a buyer alongside two colleagues. We created inquiries and orders from stock requests or direct requisitions, which our supervisor randomly distributed to us.

On return from an overseas course, I was summoned to meet the internal auditors. I was interviewed for hours on my role and how we worked. It felt like an interrogation. It was only midway through that they informed me a fraud of several million dollars had been uncovered in our unit. My supervisor had stated during his interview that I was the most active in the unit, which left the impression that I should know about such activity. I had worked for him tirelessly over the years and was disappointed that he had thrown me under a bus.

The auditors kept questioning me. I responded factually to all questions, and with full transparency. I described that there were two other buyers in the office and that random distribution of work by our supervisor meant any of the three of us could have worked on the order under scrutiny. They asked how I would know who worked on the order. I informed them that we usually signed the documents on which we worked. They handed me a copy of the documents that were the basis of the investigation. I saw the signature was not mine but that of one of my colleagues. I informed them of this, stating that this did not mean that my colleague was involved in any fraud. The approach of random distribution by our supervisor would have made it difficult for any of us to perpetuate any fraud. The auditors' approach softened after this and further conversations became friendlier.

I left the interview grateful to God for seeing me through. I reflected on why my supervisor had said the things he did in my absence. This made me suspicious that he may know more about the fraud than any of my colleagues. At the end of the investigation, I was exonerated while my departmental head was indicted.

The experience reinforced my conviction that one must remain above reproach. I learned that one needs to be careful where one works, as one's proactivity may be used by others to attempt to cover their tracks. Not all supervisors will protect one against untrue allegations, so integrity remains the best form of defence.

I learned from these experiences that it is important to be courageous and tactful, while remaining true to one's values in the face of daunting challenges. Consequences may be suffered in the process, reinforcing the lesson I had previously learned from the attack that occurred during my service year. Such consequences will be for a short time, but ultimately such a stance will produce greater benefits. As a result of my experiences, as I rose in the organization I articulated that fraudulent staff should be made to face the appropriate consequences, or they would be emboldened, demoralizing staff who upheld ethical standards.

Challenging the status quo

A practice in Shell ventures in Nigeria from the 1980s was the disposal of used company vehicles after four years to company staff through balloting, with some numbers being allocated to National Oil Company staff. This practice was intended to help increase staff mobility.

Over time, this led to open racketeering, with vehicles being sold at ridiculously low prices then sold to car dealers for a hefty profit.

By 2008, when I became VP including for logistics, I reviewed the practice against the joint venture's operating agreement (JOA). It required that any company assets disposed of should be at 'market value'. I engaged with the staff unions on the need for compliance. They raised the 'welfare' aspects and I said we could justify a 20% discount on market value in meeting the JOA requirement.

We appointed independent valuers, who demonstrated that vehicles had historically been disposed of at less than 10% of market value. The new approach and pricing were then implemented for the staff balloting and this ended the racketeering.

When this new process was forwarded to the NOC for its allocated cars, all hell broke loose. It rejected the new prices and insisted on paying the previous low prices. The NOC threatened to retaliate and withhold budget and contracts approvals if we did not revert to the former prices.

This led to a frosty relationship and was flagged by the NOC at senior levels at every opportunity. Some colleagues in Shell Nigeria's leadership, despite understanding the issues, felt it was not worth the fight, on the basis that we stood to lose more. I pushed back, highlighting that we were simply ensuring compliance with the JOA (which NOC had always insisted on) and that if the NOC preferred otherwise, they should follow due process to change the JOA. This situation went on for a while until the NOC finally agreed to the new approach and pricing.

A new leader will be faced with historical institutionalized practices that are not right or compliant with core values. This can be ascertained by reviewing practices against company procedures, shareholder agreements and best practices. Where they identify non-compliance, they should review the beneficiaries, engage them to understand any other aspects, then devise how to embed the right practice.

Pushing back against historical institutional practices can be uncomfortable, especially where there are beneficiaries of the status quo. However, a leader must have confidence and be willing to push back and do what is right. A leader must be courageous to bring change and uphold what is right.

Taking a tougher stance

In my last years in Shell Africa, I was also chairman of the Business Integrity Committee (BIC). The role covered Shell's venture companies in Nigeria, Cameroon and Gabon. I worked with auditors to investigate cases of fraud. The committee I chaired reviewed their findings and took decisions on cases.

A case in 2009 involved a staff member responsible for obtaining government permits who had submitted documents from the State Land Ministry indicating that the government had introduced a new land levy. The company was requested to pay N1.9 billion (about US$5 million) in land charges for the location of one of the company's facilities. Documents signed by a government director instructed SPDC to liaise with a state-appointed consultant to complete the process and obtain approval. Payment was to be made through the consultant to government. This was successfully processed, as it all seemed legitimate, although the charge was viewed as high.

When the excessive levy was brought to the attention of the state Governor, he said he was not aware of any such policy. He queried the land director, who absconded. Investigation by Shell auditors revealed collusion between the staff, the director and the consultant. BIC decided that the case was egregious and recommended that staff be dismissed and handed to the Economic and Financial Crimes Commission (EFCC), which recovered over N300 million (US$1 million) and prosecuted them in court.[1]

Another example occurred in 2013. Executives of NLNG Staff Cooperative sent me a complaint on their past executive's mismanagement of over N2 billion (more than US$5 million) on land acquisition. I directed an internal investigation, which established misappropriation. I authorized the complaint to be lodged with the EFCC, which recovered N200 million (US$600,000) from five staff members. I authorized the termination of their employment and commencement of their prosecution in the courts.[2]

There were efforts to delay their termination. Some in my management wanted us to merely suspend them until the court decision. Knowing how long cases can take in the Nigerian courts, this would have sent the wrong message. We had enough evidence of their misdoing, they had admitted to the fraud and they had refunded money. NLNG policy did not require court conviction to terminate staff for egregious offences. Their lawyers also threatened to take NLNG to court if we acted before court conviction. I instructed that the employment of the indicted masterminds be terminated, while others who were not directly involved but benefited from their actions were suspended without pay for several months.

Handing staff to security agents to recover stolen monies and prosecute in court is not a pleasant decision to make, but for egregious cases it enables the company to recover some of the stolen funds and send a clear signal as a deterrent. Through these experiences, I learned of the pressures put on leaders in developing countries to take a soft position on corruption. Some of these pressures are due to reputation concerns, as organizations do not like to be seen publicly as having employed fraudulent staff or as

---

[1] See https://efccnigeria.org/efcc/news/1346-n1-9bn-shell-gbaran-ubie-gas-project-fraud-how-adabanya-laundered-project-funds-witness

[2] See http://efccnigeria.org/efcc/news/1364-efcc-arraigns-ex-nlng-cooperative-chief-four-others-for-n207m-fraud

not having robust processes to prevent fraud. Some of the pressures are due to emotions, as these are staff known to leaders. In some cases, the staff unions even plead for them. Taking a strong stance when core values are violated is important, or such values will be eroded over time.

Multinationals in high-risk countries

Some major differences between high-risk and low-risk countries are weakness of institutions and a high perception of corruption in the wider societies of high-risk countries.[3] This leads to differences in societal definitions of what is right, which then blurs the integrity barometer of staff. Many view what is right more through the lens of what they see in their society, than from the perspective of a multinational's policy. For example, nepotism is common in some countries, where it is normal to give jobs to family members instead of hiring based on candidates' merits.[4] This can be normalized by local staff, contrary to a multinational's requirement for a merit-driven process.

If the same approach is used to deal with fraud in ventures of multinationals in high-risk countries as in low risk countries, it is most likely to be ineffective. An approach in Nigeria is to release indicted staff and pay them off. The rationale is that it takes time to prosecute in the local courts and that staff may institute a wrongful dismissal case against the company. As a consequence, this approach is viewed as a generous 'handshake'.

Tougher actions are needed in local ventures in high-risk countries, due to the scale of fraud in such societies.

There are examples where multinationals have taken a tough stance, such as the prosecution of a syndicated theft of US$200 million worth of oil from a plant in Bukom, Singapore[5] and a

---

[3] See www.transparency.org/en/#
[4] See https://en.wikipedia.org/wiki/Nepotism
[5] See www.straitstimes.com/business/206m-worth-of-oil-stolen-from-shells-site-in-pulau-bukom

criminal complaint against an expatriate ex-staff member in the US$390 million sale of oil field in Nigeria.[6] Such actions can make a difference.

Multinational western culture and policies make it challenging to institutionalize tough actions. I once suggested that appointment to senior or sensitive roles in high-risk countries should require staff to voluntarily declare their assets, and permit the company to verify them. Many countries have transparency laws that require this of public officials (e.g. Nigeria's Code of Conduct law). Local ventures will be acting in accordance with the law and not infringing individuals' rights. An employee nominated to senior leadership should have no concerns with declaring their assets if they have nothing to hide, and where they do have things to hide, they should not be appointed as leaders. To role-model this, I voluntarily declared my assets to my director in 2018, and empowered the company to verify them. Another suggestion was to incentivize whistleblowers when the information they provide is ascertained as factual through investigation. This would encourage more reporting of fraud. Such an approach had been used successfully in the United States for years.[7]

The company reviewed these suggestions. The former was viewed as intruding into the personal life of staff, which the company was not comfortable with, and there was concern that it could expose the company. It was feared that the latter suggestion could lead to a flurry of fictitious reports. As a result, neither of these ideas was accepted. Such challenges regarding institutionalizing tougher actions leave multinationals at risk of huge corruption in high-risk countries. I sense that the case for change for these tougher actions may also be affected by the iceberg effect, with the visible quantum of losses being seen in those high-risk countries (the top of the iceberg) clearly a fraction of the full losses beneath the surface.

In my various senior roles, I regularly received requests from officials, friends and relatives for employment, contracts and fund

---

[6] See https://uk.reuters.com/article/us-nigeria-shell-court/shell-targets-former-executive-in-nigeria-graft-complaint-idUKKBN1H415K

[7] https://en.wikipedia.org/wiki/False_Claims_Act

placements. I always encouraged them to follow due process and told them I would ensure fair treatment based on merit. I am always pleased whenever I meet people who testify to the impact of acting on merit. A medical doctor we recruited once visited my office as part of his induction. He described how he had applied for the role in Jos (900 kilometres away) and followed the process and tests. He was shocked to have gained employment, as he did not know anyone in the company or in government. 'I did not believe that was possible in Nigeria,' he said.

My ethics were strengthened throughout my career by a strong alignment between my personal values and the company's core values. However, a difficult dilemma arose as a result of an oil licence deal. In 2011, the company paid licence fees to the Nigerian Government. It indicated then that it was not aware of what the government planned to do with the income. It is indeed not a company's role to pry into how shareholders distribute their revenues, so I defended the position. However, in the courts in Italy during 2017, some evidence emerged,[8] highlighting that company officials may have been aware that part of the fees would be used by the Nigeria government to pay individuals, including a former Oil Minister.[9] It cast doubt on information previously shared to staff and undermined trust.[10] I was so disappointed at this apparent misalignment of values that I started to consider my future beyond the company.

Transparency of government finances

Nigeria ranked 146th of 180 in Transparency International's Corruption Index in 2019[11] – that is, the world's 34th most corrupt

---

[8] See   www.independent.co.uk/news/business/news/shell-s-top-bosses-knew-money-1-3bn-nigerian-oil-deal-would-go-convicted-money-launderer-emails-reveal-a7676746.html

[9] See www.bbc.co.uk/news/world-europe-42427283

[10] See www.premiumtimesng.com/news/headlines/228524-malabu-scandal-telling-lies-years-shell-admits-knew-etete-benefit-1-1-billion.html

[11] See www.transparency.org/cpi2019

nation. Publicly available reports estimate that hundreds of billions of dollars have been stolen from Nigeria since its independence.[12]

Nigeria's Statistics Bureau reports that corruption pervades the public and private sectors due to greed, weak institutions, limited transparency, weak consequences and general acceptance by citizens.[13] Governors, ministers and a police chief have been imprisoned on fraud charges. In the private sector, bank MD/ CEOs, senior company officials and business people have also been jailed for fraud.

The US Justice Department has fined multinationals for involvement in corruption in Nigeria. Technip was fined US$240 million, Kellogg Brown and Root US$402 million and Halliburton US$177 million for paying US$180 million in bribes to Nigeria officials. Siemens was fined US$800 million for bribery of officials including in Nigeria. Panalpina was fined US$82 million relating to bribery of Nigeria customs officials.

Corruption has led to high poverty levels, with the UN Poverty Index reporting the poverty rate at 80% in the north. It is responsible for poor infrastructure – for example, only 4000 MW electricity, or below 20% of what the country should have, and poor roads and rail services. It is a reason why the country's three refineries produce only 25–40% of capacity, which has led to importing petrol and diesel that is subsidized by government (10–20% of budget), a scheme fraught with corruption.[14]

One of the concerns on corruption since oil was discovered in Nigeria in 1956 has been the lack of financial transparency in the industry and a lack of accountability by government. Opaque government revenue enabled by a fragmented receipt system was in place until 2015, when a unified structure (Treasury Single Account), which had been initiated by the previous administration,

---

[12] See www.pmnewsnigeria.com/2015/01/29/400-billion-looted-in-39-years-un-official-speaks-on-negative-impact-of-nigerias-corruption

[13] See www.sunnewsonline.com/nbs-corruption-survey-report-2019

[14] See www.theguardian.com/world/2012/apr/19/nigeria-fuel-subsidy-scheme-corruption

was implemented by the new government. Before that time, over 10,000[15] bank accounts existed where government revenues were collected. Nigeria President Buhari said, 'If we do not kill corruption, corruption will kill us.' Multinationals operating under these conditions therefore face unique challenges.

I woke up one hazy morning in December 2013. I had spent the night at our Finima residential estate on Bonny Island. The Finance Minister, Ngozi Iweala-Okonjo, was visiting our plant that morning. It was her first visit to the company, and I was keen that we make a good impression.

A quick glance at my phone, which I had set to silent mode before I slept the previous night, showed an unusual number of missed calls from the then NOC's top brass. The high number of calls suggested he was frantic. Had an issue arisen overnight in the company that I was not yet aware of? Before calling him back, I decided to check with my management team, so that if there was an issue I would be armed with the most up-to-date information. There were none.

I called back and he asked whether I was expecting the Finance Minister. I confirmed that I was in Bonny awaiting her arrival. I reminded him that I had earlier notified him of the visit. He asked whether the Oil Minister had ever visited us. She had not, despite many invitations for her to visit the plant, including to commission our airstrip in Bonny.

'Can you cancel the visit of the Finance Minister?' he asked. I replied, 'It is no longer possible as the Minister has already left Abuja on her way to Bonny. Is there an issue?' He avoided the question and said he would get back to me. I wondered whether the Oil Minister was uncomfortable with her colleague's visit.

The only sensitivity I could think of was the US$13 billion worth of dividends that NLNG had paid over the years to the NOC account, outside the Federation account. The amounts paid and the accounts had been shrouded in secrecy over the years. Details were known

---

[15] See   www.researchgate.net/publication/319406328_TREASURY_ SINGLE_ACCOUNT_TSA_IN_NIGERIA_A_Theoretical_Perspective

in the presidency, by oil ministers and by the NOC. On occasion, committees in the National Assembly (Senate and House) summoned me to provide those details. I would discuss the sensitivity with the NOC GMDs (three during my tenure), but as the law empowered legislatures to seek such details, I always provided them.

Shortly after the call, I met the Finance Minister at our airstrip and introduced her to my management team. My suspicions were quickly confirmed. As we were getting into the vehicle to take her to the plant, Ngozi pulled me aside and asked, 'Babs, has your [Oil] Minister called you?'

I shared the details of my earlier call. She told me that she had been called that morning from the presidency with a request to cancel the visit. She responded that she had promised me that she was coming to visit the plant. Considering how well NLNG was doing and economic impact that it was having on the economy, she deemed it appropriate to visit.

We suspected sensitivities and agreed not to openly discuss dividends. We also agreed not to have press coverage. But I insisted I would hand her details of the dividends and financials since our inception. I felt it was right for the Finance Minister to have details of these payments. I wanted her to appreciate the size of our impact on the nation's finances. I felt that with such a picture, she would be more supportive whenever issues of NLNG came to the attention of government, such as during challenges with a government agency, NIMASA.

Her visit was a success. She toured the plant and was impressed by the standard. She commended the management, consisting entirely of Nigerians, and encouraged us to continue to show what Nigerians were capable of. At the end of the visit, while alone with her in the car to the airport, I handed her a copy of our financial statements and details of dividends. I was incredibly grateful and thanked her for the visit.

I reflected on this after the visit. We needed to make our payments to government transparent, to avoid the cloak-and-dagger situation we often found ourselves in. We had nothing to hide and our payments had been verified by the Nigerian

Extractive Transparency Initiative (NEITI), which reported that, 'NNPC acknowledged remittances from NLNG to US$13 billion, but NNPC never remitted same to the federation account.'[16]

Over the years, NLNG had been walking a tightrope to avoid upsetting government officials who did not want payments to be public knowledge. I was determined to publish our financial statements and show how much had been paid to stakeholders. Transparency would ensure NLNG would no longer be 'hushed' whenever details of government payments were required by the public.

In April 2016, I published the financial details from 1999 up until 2015 in full.[17] I used the same format that IOCs use for similar purposes globally and disclosed the full details on all income, expenditure and payments.

This was a first in the history of NLNG. It was a major milestone for transparency and the veil had been lifted. Future publications would include updated details for later years. This ensured NLNG payments has become transparent and the public now had easy access to full information, open to scrutiny and questions.

Such transparency of government payments in developing countries is important. With opaqueness, government ministries and agencies sometimes become creative and aggressive in seeking to generate revenues through unorthodox means and in many instances by using proxies.

An example during 2014–15 was the Federal Ministry of Trade and Investment (MOTI). The ministry was responsible for issuing quarterly export permits to enable companies to ship their products overseas. The minister was the sole authority for signing the permit. Before obtaining the permit, the ministry had come up with new levies for weights and measurements, inclusive of a 'metering fee' to be based on the value of oil and gas being exported.

---

[16] See https://punchng.com/neiti-seeks-probe-of-15-8bn-nlng-dividends
[17] See www.nlng.com/Media-Center/Publications/2016 Facts and Figures on NLNG.pdf

Along with other IOCs, NLNG disputed the levy. We found it inappropriate for many reasons. First, the NLNG Act granted NLNG an evergreen exemption from payment of such levies and charges.

Second, there were other government agencies (the Department of Petroleum Resources (DPR) and Customs) that had already been carrying out the same measurement of exported gas volumes for years at our facilities. While the overlap of multiple government agencies carrying out same task was inefficient, none of the other agencies had ever charged fees. NLNG had no issues with MOTI joining the other government agents, but not with it charging fees.

Third, MOTI appointed a private firm as its agent for the measurements, and for the agent's commission to be deducted from the payments. This was a red flag, as many such intermediary arrangements had been discovered globally to be fronts for officials, and had led to fines for companies.

Finally, the levy that was introduced by a regulation of the ministry was not aligned with the Weights and Measures Act, having been set as a percentage of the export value. There was no relationship between volume-measuring activities and the value of the export product. This fee was simply inappropriate.

These points were explained in correspondences and meetings with MOTI officials. Despite this, the company experienced delays over the quarterly application for the permit. They were required to either pay the levy or be denied the export permit. The objective was more about collecting the levy rather than accurate measurement.

The intensity of the demands and threat to deny the permit became stronger as the 2015 elections drew closer. I kept having to speak to the minister to remind him why we could not make such payments and the need for the export permits to be signed. I highlighted that we would have to shut down the plant if we did not get the permit, as we could not make payments that we considered illegal.

The spat continued every quarter and in most cases the permits were only issued on the very last day, late at night. On these occasions, NLNG had started to plan to shut down the plant by midnight. I always reminded the minister that, as the ministry was responsible for trade and investment, it should be attracting foreign investment. It would be a travesty for the same ministry to be responsible for stopping the exports of a company that was responsible for 14% of Nigeria's revenue. This was typical of agencies during this period.

With a new government and the appointment of a new minister in late 2015, this demand ceased and I did not need to speak with the new minister for our permits to be issued in a timely manner.

It is important for companies to be clear on the lines that they will not cross and beyond which they will commence termination of operations. They should have systems to ensure consistency of decisions in such instances. In NLNG, the shareholders (comprising IOCs and NOC) and the board are elements of such a system. The Legal Committee, consisting of NLNG's legal team and shareholder legal teams, is another element, which provides input to the CEO ahead of meetings with shareholders and the board. However, the seniority, character and disposition of the members of these systems (board, shareholders, committees) are important elements, and should regularly be assessed to ensure robustness.

Society is built on the character of citizens and organizations. Staying true to the value of honesty can be challenging in developing countries. One will regularly be tested. If one holds firm, even at personal risk, then can it truly be a value. Playing it straight can be viewed as odd in Nigeria, as many others have acted differently. Many friends and relatives stopped relating with me because they were not benefiting from my position. This was emotionally distressing, but I would not act against my ethics and company policy. To do otherwise would be wrong.

A company that is not run on sound business principles will not survive on the long run. Explaining this is often a burden for leaders, especially in a country like Nigeria. Taking such a position is viewed as a grave mistake and a lost opportunity. Some people asked, 'Why on your watch? Why can't you behave like the rest?' These type of false dilemmas and fallacies are created by acquaintances of leaders in high-risk countries, intentionally to try to force such leaders to make wrong choices. The ethical challenges for leaders in a developing country like Nigeria are enormous, but for the country to become better, all officials must do the right thing. At a church event in Aberdeen in 2014, a member described interacting with my colleagues who referred to me as 'Jesus of Shell' in recognition of my integrity.

Chapter 8

Courage

You develop courage by surviving difficult times and challenging adversity.

— *Epicurus (340–270BC)*

Staying the course

It is 6.00 am as I arrive my office in The Hague on 2 October 2017. It is earlier than usual. I could not sleep for long as my mind is 6500 kilometres away in Nigeria, from where I am awaiting news. The Federal High Court in Lagos will deliver its judgment on the case we instituted four years ago while I was NLNG's MD.

I flash back to May 2013 when a government agency used a gunship to block our exports. We had refused to pay a US$140 million levy the agency was bent on getting from us because we saw it as illegal. After weeks of the blockade, against court orders, we paid under protest to enable our export to resume. We approached the court for redress. The agency used technicalities and multiple appeals to the Appeal Court and Supreme Court to delay judgment.

My phone keeps ringing, but it is mainly team members in other countries seeking my guidance on other issues. I pace anxiously back and forth all morning, regularly checking my phone screen, hoping for a Nigeria caller ID. It is midday and my phone rings again. This time it is a caller from Nigeria; the judge has delivered judgment. My anticipation is palpable. 'We won!' said Tale, one of the NLNG lawyers. The judge ruled that our position was correct and that NLNG was not liable to pay the levy. He directed the government to refund the US$140 million collected from us and pay another US$300 million that we lost during the blockade.[1] I am ecstatic and feel relief at this landmark judgment. The courage to stand up against the government agency has paid off. Only when it is dark enough, is it possible to see the stars.

Courage is the willingness to confront danger or intimidation. It is the ability to act in the right way even in the face of threats or popular opposition. Courage does not mean absence of fear, but rather the capacity to overcome it and do the right thing. It requires one's ability to ascertain and stand with what is right and to have faith in a higher power to make things work out. It comes with maintaining a positive attitude and visualizing a favourable outcome. It is the willingness to make decisions and be ready to face the consequences. This is not always easy, especially in a developing and high-risk country, where personal loss can be real and fatal when confronting issues or interests. Courage is required in any leader determined to do the right thing.

Looking down the barrel of a gun

In 2013, a Nigerian government agency, the Nigerian Maritime Administration and Safety Agency (NIMASA), used a gunboat to forcefully block the Bight of Bonny channel, which connects Bonny River with the Atlantic Ocean. As this is the route that our ships sail through to reach the international ports of our customers, this blockade effectively halted all NLNG's production and export activities.[2]

NIMASA was set up in 1987 to regulate shipping and maritime labour. In 2007, a NIMASA Act came into effect, empowering it to collect fees and levies. The law seemed to conflict with an existing law, the NLNG Decree of 1993 (adopted as an Act of Parliament in 1999). The NLNG Act was promulgated as an incentive to

---

[1] See www.vanguardngr.com/2017/10/court-declares-nimasa-levies-nlng-illegal

[2] See https://sweetcrudereports.com/nimasa-blockades-nlng-again-over-levies

encourage the foreign investment needed to establish NLNG. It included a ten-year tax exemption and an evergreen waiver of fees and levies.

Attempts by NIMASA to collect levies from NLNG were resisted as NLNG relied on the exemption clause in the NLNG Act. Mediation by the National Assembly was unsuccessful and NIMASA went to court.

In early 2013, NIMASA withdrew the case from court. I reviewed this with my NLNG legal team and we concluded the withdrawal was because the agency realized it could not win the case. This was based on the fact that another agency, Niger Delta Development Commission (NDDC), with a similar act (NDDC Act, 2000), had lost a similar case in the Federal High Court in 2007 and an appeal to the Supreme Court in 2011 was dismissed.

We were wrong. One Friday in May 2013, a gunboat was positioned on our channel. I was informed by NIMASA that it had blocked our ships from sailing. We could not export our products. They claimed we owed US$140 million in levies. It was then I realized the agency had withdrawn the case from court to take its own action.

By evening I received a call from someone who identified himself as a prominent Niger Delta militant leader. He stated that it was his company's gunboat that blocked our channel on behalf of NIMASA, and that blockade would continue until payment was made. I listened but did not enter any discussion as NLNG had no relationship with his company.

NIMASA's action had grave implications for both us and the government. First, NLNG accounted for 14% of the government's revenue and the action would reduce the company's revenue and affect government income. Second, there were ramifications for our international reputation as our products are exported to countries that use them for electricity and heating. Third, as the government was actively seeking foreign investment to develop infrastructure, the action would be viewed as not being conducive to this, putting Nigeria's foreign investment goals at risk.

I sought the intervention of the Petroleum Minister, being the most senior government official working with NLNG. The minister requested that I meet the head of NIMASA. I flew that night into Lagos with my deputy and the shipping manager. We met him and three other agency officials at midnight, in the first-floor conference room of a hotel on Victoria Island.

During introductions, a thought flashed through my mind. This was my first time meeting the head of NIMASA, even though I had been in my role for a year. The previous court case was reason for the arm's-length approach, but I still thought I should have met with him. We could have built a relationship despite the case. I should not have had to be meeting him for the first time like this.

I clarified our position. By law (the NLNG Act), we were exempted from paying the levies. I suggested that the court was the right place to resolve any legal disagreement. I pointed to the implications of NIMASA's actions, highlighting that it was a government agency, and that the same government owned 49% of NLNG.

NIMASA's head insisted we pay since the NIMASA Act was more recent than the NLNG Act. I reminded him of the Supreme Court judgment on NDDC, stating that the NLNG Act was a Special Act and had to be specifically mentioned in any later Act for it to be impacted. Despite hours of discussion, we could not reach agreement.

I spoke with the minister after the meeting. I highlighted that if the blockade were not removed the next day, we would shut down our plant. With our storage tanks getting full, we could not continue producing if we could not export the products. She promised to follow up.

The next day, the NIMASA head called to inform me that they were lifting the blockade. He stated that this was conditioned on both parties (NLNG/NIMASA) agreeing to a mediation with senior officials. The gunboat was removed after 48 hours of the blockade and we did not need to shut down the plant on this occasion.

I convened a meeting of NLNG shareholders and the board to discuss strategy. We reviewed opinions of legal teams of shareholder companies and discussed the risks. We agreed that attending the mediation was appropriate as it would enable us to state our position to 'independent' arbiters and achieve a positive outcome.

Mediation was led by the National Security Adviser (NSA) and a representative of the Justice Minister. We had previously met the Justice Minister, who was aware of the issues and hinted that our position was strong. After introductions, the NSA announced that he had a copy of the legal opinion from the Justice Minister, which would be the mediation outcome. He wanted to read it.

I objected that it was unusual to deliver judgment without first listening to the parties involved. I stated that we were under the impression that this would be a fair mediation, where the various parties would state their positions and submit evidence, for the consideration of the panel.

The NSA reiterated that the Justice Minister had already reached an opinion, which was to be read. I again objected that it would be irregular not to hear from the parties in an arbitration. If that was the plan, he should simply have sent us the opinion by post, without the need for us to travel from our locations to Abuja. I insisted that we both needed to state our cases and be listened to before any opinion was formed and the decision read.

The Justice Minister's representative nudged the NSA and whispered to him. They then asked us to make our cases in five minutes each. This was bizarre, as we had come with a ten-page brief and evidence such as the assurance letters that had been issued by the previous Finance Minister and the Central Bank Governor, affirming the sanctity of the NLNG Act. I noticed that the NIMASA officials were unperturbed. It seemed to be a setup, not a mediation.

I made our case quickly. I cited the relevant sections of NLNG Act and highlighted that exemptions were what attracted billions of dollars. I stated that it would be a travesty if mediation, with the Justice Ministry representative present, did not respect Nigerian laws. I pointed out that as Supreme Court had ruled on similar case with NDDC, mediation should not be contradicting that court judgment. I encouraged them to read the ten-page brief and evidence we had brought before coming to any conclusion.

The NIMASA head again stated that the NIMASA law was written after NLNG Act so should supersede it. It was the same argument that the NDDC used in the Supreme Court – where it had lost the case.

Our position must have made an impression as the NSA decided he could no longer read the prepared decision. They would go back and review our positions and evidence. We would need to reconvene later for the outcome. While our protest halted the reading of the pre-planned decision, I was sure the outcome had already been predetermined.

NLNG shareholders and board were informed of what had transpired at the 'mediation'. I shared my view that the government officials and NIMASA were acting out a planned script. I believed they would come back with a decision that we needed to pay the fees.

Once the panel asked us to pay, we would be expected to do so immediately. If not, they would block the channel and disrupt our operations for an even longer period. They would then claim that we had gone through a mediation and lost.

However, I pointed out that if we also went ahead to pay the fees, against our position that these were illegal, it could place us at risk internationally. It would essentially be viewed as us making illegal payments and there were precedents of other companies being fined for such action. I was determined not to put the company at risk in this way.

A week later, on 17 May 2013, we returned to the NSA's office. He stated that they agreed with our position, that the NLNG Act exempted us from levies, duties and fees. However, he stated what NIMASA was demanding was not a levy, duty or fees, but rather a tax. He said that since the ten-year tax exemption in the NLNG Act had expired, the panel's decision was that our position on exemption was not relevant, and that we should pay the 'NIMASA tax'.

To call the NIMASA levy a 'tax' rather than a levy was curious. The ten-year tax break in the NLNG Act was clearly on company income tax, not any other tax. All fees, levies and charges were exempted on an evergreen basis. Even the NIMASA Act had never called what they were demanding a tax.

The NSA stated that the 'tax' had to be paid retrospectively from 2009 and that it totalled US$140 million. He offered us the opportunity to pay in instalments and said we should propose how we would phase the payments.

He then sent us a letter on 22 May, where he stated that the important factor in their decision was 'overriding public policy or public interest consideration' and that this was 'in the spirit of collective national/business interest and continuous safety of Nigeria's territorial waters'. These were clearly not germane, but rather mere clutching at straws and coupling of unrelated issues, and an attempt to pass off such as justification for what was clearly a biased and flawed decision.

I informed NLNG's shareholders and board of the outcome of the 'mediation' and highlighted the options. The first was to discontinue the mediation and take the case back to court. However, the risk was that it would irritate the agency and the government, and they would likely respond by blocking our channel again.

The second was to make payments in line with the outcome. This would guarantee that our operations would not be disrupted. However, as the mediation was not strictly legal, we risked international sanctions. In addition, NLNG risked being overrun by other agencies, which would simply point to the precedent that had been set by NIMASA or adopt the same brute tactics to force us to make more illegal payments.

I recommended a middle ground: that we make an initial part-payment 'under protest' then head to court. This would meet the phasing that the NSA had suggested, which should placate them. We would then, in parallel, take the case to court to protect ourselves against the other risks.

This may not guarantee that they would not block us, but it gave us a balanced approach. We discussed mitigation. One way was to seek the protection of the court against any blockade while the case was pending. The other was to enlist stakeholders to speak to the government. We also reviewed operational strategies, including accelerating the loading of vessels and strategies to manage production.

Following review with our legal teams, I proposed that we pay US$20 million (of US$140 million) 'under protest' then file a legal complaint in court. The shareholders and board agreed. We paid and informed the NSA and NIMASA, and said that we would respond later to their request to phase the balance. They were pleased.

By the next day, we had sought a restraining order. The court acceded and prohibited the government and the company who owned the gunboat from blocking our channel, pending the outcome of the case. We served the court notice on the parties immediately, including the Justice Minister (on behalf of government agencies) and the company that owned the gunboat. We expected them to obey.

However, on 21 June 2013 the gunship reappeared. On 24 June we informed the judge of the blockade, despite the restraining order he had issued. He issued a new order that the government and its agencies should vacate the channels pending the court case. They refused to obey.

We pursued our strategy to bring political influence to bear by bringing in key foreign government officials and investors to speak to the government. Ambassadors from the United Kingdom, Holland, Italy and France, and the CEOs of Shell, Total and Eni (NLNG shareholders) engaged government at the highest level. The Petroleum Minister and NOC Group Managing Director also engaged the presidency. However, the blockade remained.

With my management team, we tried to find creative operational solutions. We attempted to use third-party vessels to export our product, as they had no dispute with NIMASA over fees. The plan was frustrated by NIMASA, which insisted that no vessels be allowed to sail into the Bonny Channel.

By the second week, there was no sign that the gunship would be removed. The government was either unwilling or unable to intervene with the agency and the militant. A government directive had empowered agencies like NIMASA to retain 75% of revenues they collect, remitting only 25% to government accounts. The rational was that such agencies (e.g. Customs, the Central Bank), were classified as 'revenue generating'. The 75% of revenue retained was to be used for their expenses (capital and recurring), to enable them function speedily and not be subjected to bureaucratic delays of government budgeting cycles.

The directive meant that $105 million of the $140 million being sought would be retained by NIMASA. It would fall outside any oversight by the National Economic Council and state governments. This was the incentive, as evidenced by some of the NIMASA officials later being convicted for fraud and sentenced to prison.[3]

In the second week of the blockade, I received a call from the Secretary of the Government of the Federation. He mentioned that the issue had become an embarrassment for the government and requested that we make another payment to unlock the issues. I informed him that we had been working to find a solution. I requested that he also speak with the Head of NIMASA to persuade the agency to cooperate with us in implementing a solution.

NIMASA remained belligerent. We were losing hundreds of millions of dollars across the value chain. Some 70% of it would have gone to government, but the key players did not seem to care, as such payment would have been to the federation account and subjected to transparent oversight by all tiers of government, unlike retained revenue in agencies, where just a few people decided its use.

I reviewed our strategy with shareholders and board: our plan had not worked. NIMASA had refused to obey the court order. We agreed to pay the full amount 'under protest' while continuing our case in court.

To enable payment, we needed the court's permission, as it required a variation of the judge's earlier order. We informed the judge that it was the only way we could see for NIMASA to remove the gunboat from our channel. He granted our request on 12 July.

---

[3] See www.vanguardngr.com/2019/06/n136m-fraud-ex-nimasa-dg-jailed-7-years

We immediately paid the reconciled balance of $117 million and continued our case.

It was not until the following afternoon that we were finally able to contact the Head of NIMASA in the creeks of the Niger Delta. He then instructed the company that owned the gunboat to remove it and lift the blockade.[4]

This was background to the judgment of which I had just been informed by telephone. Our stand that payment was illegal, and that the intention behind brute action was not pure, were borne out and affirmed by the court.

NIMASA appealed the judgment on the basis that its counterclaim was not taken into consideration. In March 2019, the appeal court ordered that the case be re-heard in another High Court and judgment is awaited.

It is sometimes lonely at the top, and the blockade was an example of that. The public was not sympathetic. Some people did not understand while others benefited. It was the support of shareholders, the board and management that helped me through. The courage and strong stance of NLNG set a precedent. The judgment should ensure no other company in Nigeria will again be subjected to brute force by a government agency.

Leaders sometimes grapple with whether a different decision would have led to a better outcome. We could have selected an option that gave more recognition to the trouble that the seemingly innocuous agency could become. Considering the nexus between the agency, militants and government officials, as well as the incentive of retained income, we could have made full payment before heading to court. The option was reviewed with shareholders and the board, and the risks of being overrun by other agencies and international backlash were considered. In some dilemmas, there are no risk-free options. However, one must be courageous enough to pull back and seek a different path

---

[4] To my team members – Basheer Koko, Edith Unuigbe, Temi Okesanjo and Eromosele Inegbedion – my gratitude for their untiring work with me during the period.

when significant challenges become protracted. Finding the right balance is an art, not a science.

In leading a company, especially in a high-risk country, one is confronted with situations that would confound international management gurus. A leader must be ready to stand alone, to do what is right.

Standing firm on the law

Another challenge that I had to confront with courage while at NLNG concerned payments to government.

By 2016, NLNG had earned US$95 billion since 1999 (US$40 billion during my tenure); of this, US$33 billion had been paid to government, with US$15 billion of this money going into NNPC accounts. It was inferred that it was being 'kept aside' to enable the country to 'save' and invest in building new LNG plants (Brass, Olokola). This was not a bad idea, but such secrecy risks funds being diverted.

The remaining US$18 billion payments, mainly for purchase of natural gas (NLNG's raw material) were paid into the central federation account. In June 2014, the first payment for company income and education tax for 2013 was paid to the central account. This marked the end of ten-year tax holiday. A total of US$1.4 billon was paid, which generated public interest because it was the highest amount of company income tax (CIT) ever paid by a company in Nigeria.

The next year, 2015, was election year, and with stiff challenge from the opposition party a huge national campaign was mounted. In January, a month before the scheduled election, I was invited to Abuja to meet the Central Bank Governor and the National Oil Company's Group Executive Director (GED) for Finance. They informed me that the government was in need of revenue to meet some obligations and they requested that we immediately pay our CIT for the 2014 year.

This was unexpected and was the first time we had been approached to pay in advance (i.e. five months in advance of June, as stipulated in the CIT Act). The timing and urgency so close to an election were curious. I declined. We could not pay from accounts that had not been audited and approved by the board.

The officials tried to persuade me to pay. They proposed getting the Federal Inland Revenue Services to write a new regulation that would authorize earlier payments. I stood my ground on the basis that regulations could not override laws.

The issue was escalated to ministerial level and I repeated my case to the relevant ministers. Subsequently, there was a change in tack and the GED proposed that we pay 50% immediately and the balance in June. The desperation and urgency were concerning and the situation became tense. I informed the shareholders and the board, and they were supportive.

The pressure continued through February, but I remained unyielding. It intensified even after elections were postponed. The government had stated that the military needed time to secure some parts of the north-east, which had been seized by Boko Haram militants, so elections could be held peacefully across the country.

By April, our board had approved the audited accounts. The next day, the GED called me to request payment. He argued that the reason I had given earlier for not acceding to the request (unaudited accounts) had now been met. I reminded him of the second reason: by law we had until June.

The rescheduled elections were held in April and the ruling party lost. They began a transition to hand over to the new government in May. Yet the pressure for the payment did not abate. The departing government claimed it needed money to pay fuel subsidies, a scheme that had been fraught with fraud.

In desperation, the senior official offered that if we agreed to pay in advance of June they would credit us the interest that would have accrued had we kept the money in the bank until then. I refused to budge.

Efforts commenced to remove me as company CEO. NNPC accused me of breaching the NLNG's corporate governance protocol on LPG ship sales and demanded my removal by the board and shareholders.

The board had approved the sale of three old LNG ships. One was to be sold to a Joint Venture (JV) that had NNPC as a shareholder. They failed to make payment, in breach of sale conditions. Following the breach, we terminated the sale. In addition, a shareholder suspected that persons with links to government officials (politically exposed persons with fraud risk due to their position and influence) were involved in the JV. They requested due diligence to unveil any conflict of interest. It was the termination of the sale in line with contract terms and the due diligence that formed the basis for NOC's request that I be removed. I was not bothered: losing my job was a price I was prepared to pay for doing what was right. Thankfully, all the other shareholders insisted that I had acted properly, and the board did not accede to the NOC's demands.

A new government was sworn in on 29 May 2015 and in June (as stipulated in the CIT Act) we paid the tax and dividend, totalling US$2 billion. The new government was transparent and announced the payment. It also requested that all future dividends should be paid into the transparent Federation Account, rather than into the NNPC account. The new government distributed the payment between all three arms of government, enabling federal, state and local governments to pay the salary arrears of millions of workers. It became popularly known as a 'bailout' fund.[5] It made a positive impact across the country and enabled citizens to appreciate the huge impact that NLNG had been having on government revenues.

On many occasions, government officials, including Governors of Borno and Osun States, thanked me for the 'bailout'. They described the impact it had on their citizens. Many people who I met in towns and villages appreciated NLNG and shared how the bailout had made a difference to their lives. These testimonials of compatriots made the stressful experiences with the previous government bearable.

---

[5] See www.bbc.co.uk/news/world-africa-33435399

We had also avoided what could have been a scandal. A later revelation showed that US$2 billion was released at about same time to the NSA for arms procurement. This was to help fight insurgencies, as Boko Haram had taken control of 14 local governments in the north-east. A substantial portion of the US$2 billion was traced to having been shared among several politicians and used for the financing of elections.[6]

It takes courage to stand against the weight of a government and senior government officials, especially in a developing country. However, a leader must be willing to risk personal threats and a smear campaign – in my case, spurious allegations were levied and I was asked to be removed from the role. Doing the right thing rarely comes without a price. It is important to be clear about one's values, armed with facts and ready to engage with courage.

Being courageous requires embracing one's fears despite the natural fight or flight response. One must get out of one's comfort zone and build confidence to venture into the unknown. The key is to think through the issues and determine what is right, take a stance and be willing to put everything on the line. It is about taking risks and not being held back by fear. We all have courage inside us, but exercising it and actively reaching for it is what makes a difference. Demonstrating courage is essential for those who want to achieve the impossible.

Confronting local challenges that attract global interest

In 2011, during my time as a VP in Shell Africa there was international outrage at oil spills occurring in our operations in the Niger Delta. Some were from corroded pipes of the company that had not been replaced in a timely fashion. However, a significant majority of the spills were caused by thieves who were breaking into pipes across the company's 6000 kilometres of pipeline and

---

[6] See https://thenationonlineng.net/dasuki-lists-ex-governors-pdp-chiefs-in-2b-deals

stealing crude oil. In most instances, these theft activities led to huge oil spills that were destroying the mangroves in the Niger Delta. In addition, there were complaints about the quality of our clean-up and remediation activities.

While SPDC made an effort to get the government security agencies and communities to protect the pipelines from these thieves and prevent oil spills, the company was unsuccessful. It became obvious that some of the security agencies and communities were colluding with the thieves for pecuniary gain.[7] The theft was occurring in the public spotlight and the barges and oil tankers used were visible in broad daylight. The situation was complex and difficult to understand from a distance by international observers. As example, company efforts to quickly gain access to spill sites to stop the spill and clean-up were usually frustrated by communities wanting more compensation from a resulting higher volume of spill.

As vice president covering the environment, I decided to act quickly. There were two priorities. One was to bring transparency to what was happening. If stakeholders and international observers knew the reality, they could work together to find solutions rather than trading accusations based on misinformation. The second was to get a team of independent international observers to work with us to ensure the right quality of clean-up and remediation. Only then could we convince the world that we were serious about our responsibilities.

I worked with my team to set up a website to show details of every site where there was a spill. It showed spill location, spill date, date joint government/community/company inspection done, a copy of the signed report of joint team visit, assessment of the cause (theft/sabotage or operational), spill volume, clean-up status and photos of the spill. This ensured everyone in the world could transparently see exactly what was happening on the ground and

---

[7] See   https://ti-defence.org/wp-content/uploads/2019/05/Military-Involvement-Oil-Theft-Niger-Delta_WEB.pdf

understand the complexities. This open reporting, which started from January 2011,[8] was novel in the industry and has continued.[9]

I also met with officials of International Union for Conservation of Nature (IUCN) in Geneva in 2010. IUCN is a respected global organization in nature conservation and the sustainable use of resources. We wanted them to help assess whether our oil spill clean-up and remediation in the Niger Delta met global standards and, if not, to recommend any improvements needed.

Shell and IUCN agreed in 2011 to establish a Nigerian independent scientific panel through IUCN. It was to provide science-based assessment and recommendations for the remediation and rehabilitation of the biodiversity of the Niger Delta spill sites. This would help improve biodiversity conservation and benefit people who were dependent on the health of the Niger Delta mangroves for their livelihoods.

The IUCN panel, set up for the first time in Nigeria, worked from 2012 until 2016. It made recommendations to improve SPDC's environmental management protocols on spill response and remediation procedures. The company made changes in its approach and developed relevant guidance to protect and restore mangrove habitat. It carried out trials using biosurfactants, enzymes and sorbents to enhance the bioremediation of oil-impacted soil. Following the panel's advice, SPDC, working with Nigerian regulators, developed a framework for the management of land contamination based on international best practice. This has helped the entire oil industry in Nigeria.[10]

These two actions that I took were not all supported. Some stakeholders (including in NOC and the oil spill regulatory agency) were apprehensive about what transparency could lead to. There were concerns that it could expose a remediation gap. They were

---

[8] See January 2011 report at www.shell.com.ng/sustainability/environment/ oil-spills/january-2011.html

[9] See October 2020 report at www.shell.com.ng/sustainability/ environment/oil-spills/october-2020.html

[10] See IUCN report at https://portals.iucn.org/library/node/47915

worried it could put the security of personnel at risk. There were also IUCN members who were vehemently against cooperating with an oil company whose activities they believed conflicted with their objective of sustainable use of natural resources. Ten-years on, these initiatives have proved valuable in efforts to reduce spills, improve global understanding and improve clean-up and remediation methods.

Challenges may be of local or global proportions. During such complexities, dilemmas and uncertainties, a leader must have the determination and courage to push through improvements and seek help and support. Bringing transparency to an issue enables the enrolment of various stakeholders to find solutions, rather than trading blame. Bringing in independent international observers enables an organization in a developing country to improve its processes and national standards.

Monumental global challenge

It is 2017 and my first summer in The Hague. Since arriving nine months ago, I had enjoyed walking the 30-minutes to and from my office each day. It gave me time to reflect. I arrive in the office this June morning sweating. This is unusual. By noon the offices are quite hot and uncomfortable. There is no air-conditioning. Maybe I need a cold shower. I check my watch and the temperature reads 38°C! Later the media reports that this is the hottest summer in five decades. This is hotter than I experienced when living in southern Nigeria. It was not a scenario I had contemplated when I moved to The Hague in September 2016.

This phenomenon, and others such as the increasing devastating strength of hurricanes in Gulf Coast of the United States, is compelling evidence of the effects of climate change and human impact on the warming of the planet. There have been intensifying discussions on the climate change crisis. The United Nations' 2016 Paris Agreement set targets for countries and organizations to work towards. The long-term goal was to keep the increase in global average temperature below 2°C above pre-industrial levels and to pursue an aspiration to limit it to 1.5°C. This was to avoid a catastrophic future in which cities will become uninhabitable due to rising sea levels, drought, floods, hurricanes, food shortages, pollution, civil unrest and mass environmental migration if the temperature increases exceed those limits.

In the UN meetings in 2018 and 2019, it was agreed that the world was already heading to a tipping point, as most of the 2016 targets had not been met. Subsequently, eight climate pathway actions were outlined towards a vision of 1.5°C. The question is how fast the world can take the necessary action, as to achieve the target, the world will have to stop adding to the stock of greenhouse gases in the atmosphere by 2070.

Throughout the last century, the world has benefited from energy generated from extraction of fossil fuels. It has brought progress in transport (aviation, shipping, vehicles), electricity and industry (petrochemical), among others. It had been a key force in the world's recent progress.

However, by-products from its use – carbon dioxide (CO_2) and methane (CH_4) – contribute negatively to the climate. To achieve either of the 2°C or 1.5°C targets, the usage of fossil fuels would need to be reduced significantly or its emission eliminated.

Reducing fossil fuels usage is daunting. This is because energy demand is growing, as developing populous nations (India, Africa) require 2% annual growth in energy for development. Yet the contribution of alternative energy (e.g. hydro, solar, wind, biofuel and nuclear) is modest, accounting for only 20% of the world's current electricity need of 25,000 TW hours.[11]

Shell supplies 3% of the world's energy. It has for many years taken a leading role in the industry in acknowledging the impact of greenhouse gases and finding solutions to reduce emissions. It was a pioneer in the use of LNG in the 1970s and has advocated its increased use in place of coal. LNG is a cleaner fuel that emits

---

[11] BP Statistical Review of World Energy (2018) at www.bp.com/content/dam/bp/business-sites/en/global/corporate/pdfs/energy-economics/statistical-review/bp-stats-review-2018-full-report.pdf

only about half of the CO_2 of coal. In the 1990s, Shell invested significantly in alternative energy (solar and biofuel). Shell has also taken actions to reduce flaring in its operation (e.g. 90% flare reduction in Nigeria), thus contributing to reducing the impact of climate change.

Shell has continued to work to bring stakeholders together globally to discuss what policies, investment and research are required to achieve reduction in emissions. In 2016 it set a target of reducing the net carbon footprint (NCF) of the energy products it sells by 50% by 2050, 20% by 2035 and 2–3% by 2021 compared with 2016.[12] Shell has also included NCF reduction in its management remunerations and aims to have net zero emissions from its own operations, categories one and two, by 2050.[13]

A path was developed for emission reductions in Shell's products.[14] It includes carbon capture and storage; divesting and decommissioning high emissions assets; shifting investments from oil to gas with lower CO_2 emissions; increasing renewable business (US$2billion/year, or approximately 10% of its capital expenses per annum); increasing biofuel production; electrification; and growing natural sinks (forestation/planting trees).

Shell has made significant progress on the plan – for example, it created a new division (New Energies), which has acquired renewable businesses including First Utility (UK household energy),[15] EOLFI (French floating wind),[16] Sonnen (smart energy storage), Orb Energies (India Solar), Silicon Ranch Corporation (US solar), Greenlots (US electric vehicle charging), NewMotion

---

[12] See   www.shell.com/energy-and-innovation/the-energy-future/what-is-shells-net-carbon-footprint-ambition.html

[13] See   www.shell.com/energy-and-innovation/the-energy-future/what-is-shells-net-carbon-footprint-ambition.html

[14] See https://reports.shell.com/sustainability-report/2018

[15] See   www.shell.co.uk/media/2018-media-releases/shell-completes-acquisition-of-first-utility.html

[16] See   www.shell.com/energy-and-innovation/new-energies/new-energies-media-releases/shell-agrees-to-acquire-eolfi.html

(energy provider in Europe), Limejump (UK energy technology) and GI Energy (microgrid).

A challenge for the industry in reducing CO_2 is making carbon capture and storage (CCS) commercial (i.e. without the need for subsidy) and to scale it up a hundred-fold. It is still at R&D stage and more work is required in this area as it is key to reducing CO_2 emissions in the industry.[17]

On nature solutions (forestation), strategy is evolving. Stakeholders have been supportive of capturing CO_2 from growing or preserving trees, but it is not yet clear whether the world will accept increased forestation and its absorption of emissions as a compensation ('quid pro quo') for continued fossil fuel emissions by companies.[18]

The climate crisis is a threat to our existence and a challenge for the industry. Shell has made a good start in addressing it over the years, both on thought leadership and in concrete steps (targets, report, portfolio, investment, link to remuneration). Some have argued that Shell started late and was not strongly on the side of the climate debate early on; however, Shell had always articulated the need to move to cleaner fuels (gas) and years ago moved to lead the debate and bring stakeholders together. Knowing the calibre of the people in the industry, I am optimistic that this threat will be significantly addressed. Shell has the courage, people and ability to play a major role.

In my role in Global Shell Upstream Leadership (2016–19), as Upstream Vice President, reducing emissions across Shell's Upstream operations in 40 countries was a priority (along with the process integrity of 100+ sites and the personal safety of 100,000 staff and contractors in Upstream).

We achieved a 20% reduction in emissions across our Upstream business between 2016 and 2018. We ensured there

---

[17] See www.gov.uk/guidance/uk-carbon-capture-and-storage-government-funding-and-support
[18] See www.iucn.org/news/nature-based-solutions/202007/iucn-standard-boost-impact-nature-based-solutions-global-challenges

were greenhouse reduction plans for all producing locations. We knew all emission sources, volumes, identified and implemented strategies to reduce the emissions. For example, our shale operations in the United States changed the way we appraised and commissioned new wells and eliminated significant emissions. The company also introduced drone technology to enhance its methane leak, detection and repair program.[19] We implemented new guideline to end routine flare by 2020 and to shut down production when flared gas in assets exceeded limits. Countries like Nigeria delivered gas-gathering projects to capture flared gas. In some fields in Europe, we moved from diesel generators to electrification.

Organizations periodically face existential threats from the market, society and even government. A leader must step back to analyse the threats and engage widely to find solutions. Enrolling stakeholders and having the courage to drive through the solution is crucial. A leader must work to confront such threats, protect the organization, find value where possible and do the right thing. Oil spills and climate change were challenges of international dimensions, while transparency of government revenue and refusal to pay illegal levies were issues of national significance. These threats required significant leadership efforts and responses within and outside the organization. A leader must recognize such existential threats quickly and act expeditiously.

---

[19] See www.shell.us/media/2020-media-releases/expanding-use-of-drones-for-methane-detection.html

Chapter 9

Relationships

A growing relationship can only be nurtured by genuineness.
— Le Buscaglia (1924–98)

Good neighbourliness

My flat in the Hague is now empty; the movers packed up my belongings and are now gone. Shortly afterwards, the landlord and his agent arrived for handover formalities. My landlord is an elderly, easy-going man. I had met him in his office and law firm when I was about to start the tenancy. He has a warm disposition. As we walk around, we chat about which country I am going to next. He reminisces about the previous tenants who have left for various countries. I tell him I am returning to Nigeria, my native country, as I have now retired. He looks at me coyly, probably wondering why I would be retiring so early. We finish the inspection and I sign off the home inspection report. I hand the keys to him and we shake hands and say our goodbyes.

Downstairs, I say farewell to my neighbours, Amy and Arnold. An elderly Dutch couple, they had lived in the flat directly below mine for many years. I had grown closer to Amy over the years.

The last winter that I was in the flat, the heating system broke and my attempt to fix it was not successful. I went downstairs to consult my neighbours. Amy was sympathetic and offered to help. After fiddling with the nobs for 10 minutes she asked when I had last changed the heater's filter. I had never done so. She smiled and mentioned that she suspected the filters were clogged, affecting the efficiency of the heating system. She gave me tips on how to change the filters and invited me over for tea and biscuits while I waited for the flat to heat up. I spent an hour with Amy, and we shared some stories. Amy was keen to know about Nigeria. She had read stories on social media on Boko Haram and wanted to know how safe Nigeria was. I described the size of Nigeria and that the impacted area was less than 2% of the country. I assured her that most parts of Nigeria were safe.

They wish me well in the next phase of my life. I alight from the lift and put my bags in the boot of the waiting taxi. We begin the drive to the airport to catch a flight back to Nigeria. I take one last look through the rear windscreen at the building that has been my home for the last three years and it brings nostalgic feelings. I will carry the beautiful memories of my three years in the Hague flat like a diary.

Building relationships is key to achieving results. However, it is important to develop them before they are needed. In building relationship, one must first give. This entails sowing, so that when one needs to reap, the seed is well planted. One should first find out what is important to the other person or what they are working on and support them to achieve their objectives. Enabling others first gives one goodwill, which one can leverage in the future.

Building relationships requires developing people skills and emotional intelligence. It is important to identify the need for the relationship and schedule time to build it. One must listen and appreciate others. It is about enabling others, being inquisitive and sharing oneself. Relationships are built individually through friendly connections. They also require getting involved and participating in activities and doing things together (e.g. charity events, social events) but should not be built on unethical grounds.

Building relationships to unlock value

One of the relationships that I built to unlock value as NLNG CEO was with the Governor of the host state of NLNG. I first met the Governor, Rotimi Amaechi, following his attendance at my welcome party at the Presidential Hotel Port Harcourt in 2011.

We spent time at the event getting to know each other. We talked about formal issues (state and industry) and personal matter (family, social interests) and shared some jokes. He was upbeat that NLNG had relocated its head office to Port Harcourt, partly due to his prodding, and he assured me that he would support the company. We agreed to continue to build a close working relationship.

I began to meet Amaechi to understand his priorities and see which of them aligned with our values so I could make a difference. An example dear to him was the Model Schools. Oil companies had promised to build one each in the state, but none had completed in over three years. The challenge for the companies had been the complexities (security, community) that came with such an activity. I took this on and developed a new, more pragmatic approach supported by the board. This was premised on the state going ahead with building the school while we would pay on the basis of milestones reached. Amaechi was happy that I was able to help unlock this issue.

One challenge in relationships is the dilemma of working for a global company with international standards in a developing country like Nigeria with conflicting standards. This has the risk of creating tension in the relationship with government functionaries. In an example, Amaechi took me to his security tower with an impressive bird's-eye CCTV view of the state capital, providing the ability to identify crime and respond rapidly. He discussed his plan to acquire two helicopters for 24-hour surveillance over cities and across pipelines. He asked for support from us, along with the other IOCs, as improved pipeline surveillance would benefit us. I explained that we could not support such military grade security apparatus due to our governance processes and commitment to the Voluntary Principles on Security and Human Rights.[1] Although he was disappointed, he also understood and appreciated the clarification and frankness.

---

[1] See www.voluntaryprinciples.org/the-principles

Building relationships requires honesty, as it is not possible to agree all the time. Once I discussed with Amaechi that we were working on supporting financing the construction of the road from Port Harcourt to Bonny (the Bonny–Bodo Road) with NLNG contributing 50% of the cost. The Bonny–Bodo Road is a federal government project that had been abandoned for 40 years. He was doubtful about the federal government's commitment and suggested I instead consider the option of extending the road he had progressed to Andoni, which is next to Bonny. I promised to investigate his suggestion. A review by my team highlighted that it was a credible and competitive option, but there were political risks, especially with the elections looming. I engaged him through the review and decision process to continue with the Bodo option. The humility with which I addressed this delicate issue ensured there was no damage to our relationship. The long-term benefit of shunning hypocrisy far outweighs any short-term comfort level.

The relationship grew and he was accessible whenever I needed to discuss NLNG issues. I found him hard working, well-meaning and forthright, attributes that enabled alignment. People had various views about his prioritization, due to the mega projects (e.g. monorail) that he undertook in the face of the high level of unemployment prevailing in the state, where citizens were more in need of the basics for survival. However, he was also making progress on fundamental areas such as schools, clinics and roads.

In all our interactions, I never doubted his passion to develop the state, as he never showed interest in personal aggrandizement and was always focused on the public good. In contrast, I met other government officials whose first interests were either how to appoint their people to NLNG's board or how to divert funds from major projects that we were undertaking to finance their personal pet projects.

One key challenge during my tenure as CEO of NLNG was Train 7, a planned expansion to the existing six trains (production lines) in the company's plant. It would add 40% volume, increase revenue and offset the impact of forecast lower future gas prices.

In 2008, NLNG completed a front-end engineering design (FEED) on Train 7. However, the project stalled due to gas supply issues and expensive quotes from vendors, which truncated the final investment decision (FID). It was decided to hold the project in a 'keep warm' mode and maintain skeletal activities (technology consultancy, strategy review and minor works on site).

In 2010, the Nigerian President died and was succeeded by his Vice-President, from Bayelsa State, who appointed a Petroleum Minister from same state. This shifted the interest of the government to another LNG project to be situated in Bayelsa (Brass LNG) as a legacy for the then President and Minister in their Bayelsa State. Some argued that Brass LNG would diversify from the NLNG monopoly. While this has merit, the proposed location of Brass, on the Nun River estuary in the Bight of Benin, was not ideal. The location is significantly silted and requires regular dredging, which will significantly increase its shipping costs.

It was also not feasible for both projects to be sanctioned at the same time. Each would cost US$10 billion for plant and upstream projects for the supply of input gas. Such a level of investment in one go, in a high-risk country such as Nigeria, would not easily be financed by the common shareholders (NNPC, Total, Eni).

Both projects would also need significant input gas. While Nigeria had enough gas reservoirs – 187 trillion cubic feet (tcf) of proved reserves and 600 tcf unproven reserves – their development would take time due to the construction complexities (security, community) in the Niger Delta.

Despite Brass LNG's political support, its economics, selected technology and feasibility were challenging, as a new greenfield project – unlike Train 7, an expansion and brownfield project. Train 7 was cheaper to build and would benefit from economies of scale, as NLNG already had infrastructure in place for the previous six trains.

Three shareholders of Brass LNG (NNPC, Total, Eni) were also NLNG shareholders. Perhaps for commercial and political reasons, they continued (with ConocoPhillips, technical partner and technology provider) to work on Brass LNG, providing hope to the President and Minister.

As a result, Train 7 fell from favour and remained stuck due to the loss of government support, which is crucial for any LNG project. However, we had to move as the window of opportunity was closing. Shale gas was adding competition and a price decline was looming. NLNG needed additional volume to compensate for the price decline and to grow market share.

The pursuit of Train 7 became a pivotal priority during my tenure. Its economics were good and it ranked high in the portfolio of shareholders, including NOC. The challenge was to (re)gain political support, crucial for a project of this magnitude as government plays a big role in LNG projects worldwide.

With my team, I developed the engagement strategy. One strand was to enrol leaders with influence to sway the President and Minister, to enable us to bring shareholders (including NOC) on board. The other was to raise public awareness and clear the myths that had been released into the public domain in the past, such as the myth that there was inadequate gas for domestic and export use.

To enrol national and state leaders, we started naturally with River State, as the host of NLNG. Amaechi was in the same party as the then President. They were friends and he had the President's ear. I explained Train 7 and the value it would bring to his state (jobs, US$10 billion investment). NLNG would raise the finance itself. He was struck that there was no requirement for government to provide funds for the project despite its value.

He was enthused by the benefits to his state (jobs) and country (revenue) and could not understand why the President and Minister were not supportive. I sought his help to engage them, to enable the project to proceed. He assured me that he would take it up with them. I was pleased and expected he might have a quiet word.

Weeks later, he confirmed that, during a break at a public event, he asked them why they were not supporting Train 7 and said he needed their support. They admitted to him that they were withholding support until Brass LNG's FID was sanctioned. Amaechi was furious. Others described it as a very agitated discussion.

Despite the confrontation, he did not achieve the desired results; instead, it seemed to have hardened their position. However, it provided confirmation, as prior to this it was mere speculation that the President and Minister were not supportive. Now we knew they were impeding the project.

The disagreement about Train 7 was one of the issues that led to a breakdown in the relationship between Amaechi and the President.[2] Other issues were the dispute about ownership of oil wells between Rivers State and Bayelsa State[3] and a misunderstanding with the President's wife.[4] Amaechi quit the ruling party and joined the opposition, which later appointed him Transport Minister when it won the election.

We continued a healthy relationship beyond our tenure. In 2016, Amaechi attended my send-off party in Abuja. Relationships that are built on honest interactions and the benefit of both parties are long lasting.

Other leaders of influence with whom I built relationships were former presidents who were not members of any political party, so that their involvement would not be politicized. I invited former heads of government (General Gowon, Chief Shonekan) to visit the plant in Bonny. They had played supportive roles during their tenure to make NLNG a reality. The Gowon era (1966–75) had conceived NLNG, developed Nigeria's gas master plan and obtained the approval of the US Energy Department. The approval of the US government is a requirement for construction of plants that export products to US terminals. Chief Shonekan's interim

---

[2] See https://punchng.com/amaechi-didnt-betray-jonathan

[3] See www.herald.ng/jonathan-took-rivers-oil-wells-bayelsa-president-amaechi

[4] See www.vanguardngr.com/2015/05/where-jonathan-and-i-disagreed-amaechi

government (1994–95) was instrumental in resolving several of the constraints that had delayed the setup of NLNG.

Their visits, at separate times, provided the opportunity to share with them the progress that had been made since their efforts. It was fascinating to see the pride in their eyes as they toured the complex and visited the LNG ships. They visited King Edward and saw community projects. I later visited them at their homes on occasion to continue to build the relationship. They spoke up in public on the need to progress the Train 7 project.

Their visits and the media coverage they attracted also enabled us to correct some of the erroneous myths being circulated. One was there was not enough gas for export and domestic projects. Some 80% of electrification in Nigeria was from gas, but with only 4000 MW of power being generated, power cuts were experienced most of the time. There was thus a strong drive for more gas to be directed to electricity generation. We explained that Nigeria had 187 tcf of proven reserves and 600 tcf of unproven reserves. The total requirement for the two export projects and domestic gas was below 150 tcf; thus, technically, gas reserve availability was a non-issue.

Another myth was that Train 7 was in competition with Brass LNG. This was myopic, as competition was between Nigeria and other countries and not between projects in Nigeria. Qatar was building 77 million tonnes per annum (mtpa), Australia had 80 mtpa while all the projects in Nigeria, if built, would deliver a much lower 50 mtpa compared with the other countries. We also enabled Brass LNG to visit the NLNG plant and for their staff to spend time in our organization.

The enrolling of leaders and engagement of the then president continued through May 2015 when President Buhari was elected. Along with the board, we met the new president in 2015. He knew of NLNG, having been Commissioner (Minister) for Petroleum Resources in 1976 and Chairman of NNPC in 1977, all as part of the General Obasanjo's administration. He had become aware of the delay to Train 7 from our campaign. He said that the government had originally planned for 12 trains. He was happy with progress to six trains but disappointed that the previous government had not supported the seventh and eighth trains. He signalled to the Oil Minister and NOC GMD that they should provide support. This was a major outcome.

NLNG Board visit to President Buhari in 2015

Train 7 thereafter made quick progress, resulting in the signing of FID by the shareholders in December 2019 and an EPC contract by May 2020. The efforts to build relationships on Train 7 during the 'keep warm' phase had paid off.

Building relationships is an important role for a leader. It should occur early and with critical stakeholders to advance the objectives of the company. The efforts cultivated with President Buhari led to a more conducive environment, crucial for the Train 7 project to thrive and move forward.

Building relationship for national value

Another set of relationships built to unlock value as NLNG CEO was with National Assembly legislators in the 7th (2011) and 8th (2015) national assembly (Senate and Representatives).

I began by meeting the Senate Presidents and House Speakers. We interacted during visits to explain NLNG's contribution to the nation. They were aware of NLNG mainly from issues raised by legislators, their constituencies and in the media. Our interactions enabled us to clarify issues. Meetings were held in their offices, but also sometimes at their homes, late at night. They were busy but listened, demonstrated understanding and were keen to create a cordial working relationship.

Building umbrella relationships with principal officers was an important step ahead of relationships at committee levels. It provided a strong backdrop, critical in the hierarchical cultural context of Nigeria. I subsequently met the chairpersons of the Senate and House Committees for Oil and Gas at their offices in the National Assembly. We became quite friendly over time and I would sometimes visit their homes for social visits when I was in Abuja on official engagements.

One challenge during my tenure as NLNG CEO was the NLNG Act, promulgated as an incentive scheme to attract foreign investors to build NLNG. It was initially a decree in 1993 and later adopted as an Act of Parliament in 1999. It included a 10-year tax holiday and the provision that levies, charges and fees would not apply. Its sanctity was affirmed by Supreme Court in 2008, in a case brought by a government agency, the Niger Delta Development Commission (NDDC), which had tried to impose some levies on NLNG (see Chapter 7).

Some political stakeholders of government agencies like NNDC and NIMASA, as well as some legislators were unhappy about evergreen incentives in the Act and felt it should have been time bound. They believed IOCs were making too much money from NLNG and that Nigeria should make more. They sought to unilaterally amend the Act contrary to provision that any amendment would require the agreement of shareholders. Legislators felt that the clause on consulting shareholders before amending the Act violated their constitutional powers to unilaterally make laws. However, any unilateral action would violate the agreement that enabled investors to bring in billions of dollars to build NLNG and such action would put the country at risk of international litigation.

It was therefore imperative to defend the Act. We developed a two-pronged strategy. One approach was to engage legislators and government, to ensure they understood the implications of what they were embarking on as not being in Nigeria's best interests. Nigeria had to demonstrate that we honoured our commitments. The second was to clarify the rationale for the Act and debunk myths about its key elements.

The relationships I had built with the chairpersons of the oil and gas committees became valuable for our efforts to engage the National Assembly on the amendment of the Act. I explained the issues to them informally and formally, and they demonstrated understanding. They facilitated discussions with their committee members and clarified that the committees would need to respond to the petitions received from the government agencies and the demands of political stakeholders. We also engaged the executive arm of government to provide support. In one of the legislative public hearings, the Attorney-General and Minister of Justice presented the case for why any unilateral amendment of the Act would not be in the nation's best interest and this helped to make the legislators reconsider.

This reinforced the second part of our strategy and the need to raise public awareness. We shared with the public that NLNG had been in the design phase for 30 years. It was the Act that enabled it to take off and attract foreign investment of US$7 billion to build the first two trains, to reduce gas flaring from 65% to 20% and for Nigeria to have earned US$33 billion by 2016. The Act enabled NLNG to grow from two to six trains. Retaining the Act was crucial to giving shareholder and finance institutions confidence to fund US$10 billion required for Train 7, which would generate 40% more income for Nigeria.

The NLNG Act was not unique, as incentives to attract investments are a global practice since investors have options about where to invest. Developed countries such as the United States and United Kingdom offer more generous incentives to attract investors. Even in Nigeria, there were Free Trade Zones (e.g. Onne) where companies could enjoy evergreen incentives on levies and fees, as well as on income tax, whereas for NLNG it was capped at ten years.

The government owns 49% of NLNG and earns a commensurate dividend. NLNG also pays 30% income tax, following the expiration of its 10-year tax holiday. As a result, 70% of NLNG's profit goes to government. We therefore highlighted, as part of the awareness drive, that the highest beneficiary of the NLNG Act and incentives (financially at least) was the government.

We also set up an engagement session with CEOs of major media outlets (newspapers, televisions). Meetings were usually held as an informal quarterly breakfast session. They enabled the company to directly interact with media leaders to provide factual insights into some of the issues and discuss the benefits to the country and the risk if the issues were not resolved appropriately. We also used the sessions to answer any questions and clarify misconceptions.

Efforts put into building relationships with stakeholders enabled us to add value and protect the company from actions that would have led to losses for both the company and Nigeria. The NLNG Act was not amended during the five years I was at NLNG and no actions were taken on any other laws that undermined the NLNG Act.

A US$449 million judgment[5] by a Bilateral International Treaty Tribunal (BIT) against the Ecuadorian government to compensate an oil and gas investor (Perenco) for Ecuador's unilateral imposition of windfall tax in breach of the incentives given at the onset provides an insight into what was avoided by

---

[5] See https://cf.iisd.net/itn/2019/12/17/substantial-damages-awarded-to-perenco-for-fet-breach-and-expropriation-ecuador-also-awarded-compensation-under-environmental-counterclaim-perenco-ecuador-limited-v-ecuador-icsid-case-no-arb-08-6

efforts put in to ensure the NLNG Act was not amended against provisions in the Act. Our efforts helped Nigeria dodge a bullet.

Going beyond boundaries

From the engagements with stakeholders on issues that confronted NLNG (Train 7, the NLNG Act), it became clear the company was not sufficiently well known across Nigeria. Beyond our host state (Rivers), stakeholders had little knowledge of NLNG. We were an export company, so our main product was not used in Nigeria; citizens therefore had little or no direct contact with NLNG. Although one of our products (LPG cooking gas) was used in the country, we were only involved with its supply to the terminals. We were not involved in the retail, distribution or branding aspects, which would have enabled more interaction with the public.

Our plant was isolated on Bonny Island and only few stakeholders had knowledge of us. This became obvious during our blockade by NIMASA as only very few notable stakeholders spoke up for NLNG. Because of our size and impact, many of our issues were of national significance (e.g. government revenues, NLNG Act, NIMASA). Being limited in our interactions and exposure was therefore not to our advantage.

It became strategically important to extend our presence. We needed to build relationships more widely in order to acquire political capital beyond Rivers State. It was vital that governors, ministers, legislators and stakeholders from other parts of the country knew us so that when there were issues that required wider external support, there would be more stakeholders who could speak up for NLNG.

In 2013, I reviewed the situation with my team, and we considered our options. We looked at starting a football competition across the country. Nigerians are passionate about football – it is said to be an activity that unites all ethnic groups. We looked at intervening in the health sector, as hospitals and medical care across the country needed rehabilitation. We assessed these as good opportunities, but felt there was only a tenuous link between those options and our heritage, and we had no experience in those fields.

Over the previous decade, we had sponsored an annual Science and Literature Award for Nigerians of all ages, with a cash prize of US$100,000 each for the winning entries. The competition generated huge interest each year and entries came from within Nigerian and internationally. We concluded that education was an area of strength for which we were known. We reviewed the history of the prizes and noted that while we had winners every year for the literature prizes, there were many years when there had been no winners for the science prize. This reflected the decaying standards of universities in the country, especially for science-based research.

As an engineering company, we identified technical research as an area in which we could intervene. Improving the quality of education at university level would enable us to attract more winners for the science prize in future. We decided to build and equip modern engineering laboratories, one in each of the six geopolitical zones of the country. This would enable us to build a network bridge and improve our name recognition across Nigeria.

The proposal of spending US$2 million in each university was supported by the board. We engaged the universities and I met with the Vice Chancellors. They developed their own laboratory designs. They also decided on the laboratory equipment that they wanted, depending on the engineering discipline they chose. The only limit they had was the US$2 million budget ceiling.

To ensure value, NLNG cost engineers and procurement staff reviewed the proposals and cost of materials for the universities' building designs. They carried out global benchmarking of the prices for the equipment. Working with the universities' teams, they formulated a good approach for the construction and procurement, and sought opportunities for joint procurement between universities.

There were suggestions by NLNG staff that we carry out the design, construction and procurement of equipment. I decided that we would not use company resources for this; our staff were already stretched with our internal work. In addition, many university locations were a huge distance away, so the logistics would be challenging – especially as we did not have local knowledge of contractors and vendors in those locations. The universities were better placed to do this. Besides, our engineers were trained to build more complex facilities than those required by the schools. For example, we would normally build explosion-proof buildings in our plants, due to the risks from our products. Getting them to design and build the laboratories carried the risk that they could take the thinking and approach of oil and gas standards into the design and construction of these laboratories, with the attendant higher costs.

The approach worked. Universities in Ibadan, Ilorin, Maiduguri, Zaria, Nsukka and Port Harcourt were the first beneficiaries in 2014. They built excellent laboratories and procured modern equipment. The University of Port Harcourt procured a modular refinery (a prefabricated processing plant on skids). There was no modular refinery in Nigeria at that time. The university planned to use the facility to teach students how to operate and manage modular refineries to develop capacity in the region and create significant employment, as well as help minimize environmental damage. Ten modular refineries are now in various stages of production in the Niger Delta. Many of the students who graduated from University of Port Harcourt and were trained in operating modular refineries are being employed by those companies.

The building and equipping of engineering laboratories have enabled NLNG to have a presence in five new states where we previously had none. As part of the project flag-off ceremonies, we met their state Governors, deputies, traditional rulers, legislators and other stakeholders who had not previously had interactions with NLNG. This provided opportunities to build closer relations with more stakeholders and to engage them to become supporters of NLNG. The strategy is that, in future years, one other university in each geopolitical zone, at intervals, will be selected and similar development progressed.

For a leader to be successful and achieve the goals of the organization, they need to build and maintain healthy relations with stakeholders who may or may not be directly involved in the organization's activities, including government, legislators, special groups and the public. Developing relationships is not always easy, but it is a crucial part of a leader's role. A challenge in developing relationships in high-risk countries is how to ensure that they are built on core values that are altruistic rather than benefiting only individuals. Walking this thin line is a challenge, but it is one that must be met with clarity and effort.[6]

University of Ilorin NLNG Engineering Laboratory
Commissioning (2016)

---

[6] To my team members – Isa Inuwa, Siene Alwell-Brown, Kudo Eresia-Eke and Fola Olanubi – my thanks for their efforts, ideas and support on the several initiatives on relationship building.

Chapter 10

Reflections

No one can ride your back unless it is bent.
– Martin Luther King (1929–68)

Orderly transitions

'There is one more thing I would like to discuss, Marvin,'
I say to my boss in December 2014. We have just gone
over my appraisal. He had given me feedback and areas
to improve for the next year. 'I would like you to start the
process of identifying my successor.' Marvin was silent for
few seconds. I could hear his breath on the phone. 'Why
so soon?' he asked. Before I could respond, he followed
up and said, 'You have only spent three years in the role
and you are doing well.' I share my thinking. 'From my
perspective, a leader should not be in a role for more
than five years. The first year is coming to grips with the
job, learning and building relationships. The second year
is evolving strategy and enrolling others. The third year
is consolidating strategy and starting implementation.
The fourth and fifth years are realization and delivery of
improvements. Beyond this, there is a tendency for leader
to start cruising, listening less and becoming less sensitive
to changes around them. Normalization starts to set in.'

By mid-2015, Marvin began the search. We discussed
candidates and some were interviewed. My successor was
selected and commenced a one-year programme, visiting LNG
plants, customers and partners. He was endorsed by the board at
a meeting in London in July 2016. The handover ceremony held

in Abuja, attended by Ministers (Rotimi Amaechi of Transport and Amina Mohammed of Environment, now Deputy Secretary-General of the United Nations), Governors (Rivers and Kwara State), NLNG and BGT Board Chairmen and Directors, Brass LNG Chairman and other dignitaries. My work at NLNG done, I then proceeded to The Hague to resume my place in the Shell Global Upstream Leadership Team in September 2016.

Growth in life is about being able to reinvent oneself or to inventing a new future. In my journey from the storeroom to the boardroom, there were many elements that I had to reinvent, including focusing on others and working on my own struggles.

NLNG board send-off (2016)

Motivating others

Through my journey up the career ladder, one of the most important personal transitions was the shift from focusing on my own individual performance to focusing on enabling others to excel. During my early years in the storeroom, my focus was mainly on delivering my own tasks and driving myself to excel, but the higher I rose, the focus was less on my personal deliverables and more on deliverables of team members. The more a leader can develop, motivate and inspire staff to deliver exceptional results, the greater the results their overall organization will achieve.

I had learned from my parents how much of an impact a leader can have on others. Having studied in Nigeria and overseas, and worked in north and central Nigeria, they were welcoming of diverse people. They believed in talent wherever it came from. Many of my father's students went on to prominent positions, including a federal minister, Mr Lai Mohammed, an Editor of *Newswatch* magazine, Mr Yakubu Mohammed, a former Board Director of the Central Bank, Chief Joshua Omuya, and a Non-Executive Director of Nigeria Communications Commission, Mr Clement Baiye.

Clement acknowledged that my father was admired for several things, including his integrity. Another aspect was his inspiration of others. In 1971, my father encouraged him to participate in the John F. Kennedy (JFK) Memorial Essay Competition. He won the state prize and ranked in the top 10 nationwide. During this period, Clement was appointed 'food prefect', and there were issues over the responsibility. Clement approached my father requesting to resign. My father chided him for baulking under pressure, telling him not to walk away when provoked, and that he should continue because he had my father's confidence. Clement took the advice and subsequently excelled.

A leader is more impactful when able to demonstrate empathy and show compassion. Supporting others to succeed and working to alleviate their issues brings emotional connection. In high-octane work cultures in many companies, the macho leadership tendency to suppress emotion is unhelpful for a leader who intends to motivate and inspire staff to exceptional performance. Staff will move mountains for a leader who they believe cares for them and has their back.

A leader with a vision will achieve results with talented, competent, motivated and inspired staff. A key role of a leader is therefore identifying talent, empowering, coaching and mentoring staff. This is an area on which a leader will increasingly spend significant time as they rise to senior roles.

Having been a teacher, this came naturally. I had also learnt from my bosses, who had inspired me to better performance over the years. Helping to nurture staff and see their transformation and progress was always fulfilling and a joy. I had the privilege of coaching and mentoring hundreds of staff.

Coaching focuses on a specific here-and-now issue. I approached it by getting staff to reflect on the issue: What created it? Why? Are there any related factors? What are the options to resolve the issues, pros and cons? I avoided offering solutions but got staff to reflect on the best solutions and how to implement them. That way, solutions were more relevant.

Many staff expected answers from me as a leader, and I always had to steer them gently. A coach's role is to guide through reflection. However, in cases where I had been through a similar experience, I would share this with them. For example, to those who were feeling poorly appraised by their supervisor or not having a good relationship, I would share my stories of Warri and Aberdeen.

Mentoring is focused on long-term career. I encouraged staff to first step back and think of their values, important things in life and what they enjoy doing. I encouraged them to think of their career in terms of this context. Without situating ambition in important things, one will not find fulfilment. Where appropriate, I would share stories of my career and the things I found important, including the pre-eminence of God and family.

A recurring experience during mentoring is that staff expect it to include 'sponsoring'. While the sponsoring of talent is one role of a leader, it is not the same as mentoring. One can mentor many staff but can only 'sponsor' a few top talents. Sponsoring is about putting the right people in the right position. It is a role that a leader needs to discharge without favour as there is more at stake than sparing feelings. Putting the wrong people in a position will slow down a vision and discourage gifted staff, who will wonder how long they will last. It is key to set clear expectations at the start of a coaching or mentoring relationship.

I was always glad to see the staff I had coached and mentored go on to resolve an issue and excel in their careers. Authentic fulfilment comes from contributing to others' success.

The other important element for a leader is holding people accountable to deliver results. It can be tough, but a leader must always strive to be objective and consistent in how tasks and rewards are allocated, between staff. I always expected excellent results from staff, and was intolerant of mediocre performance and poor behaviour. Our benchmark was international as we worked for a multinational, and we were all well remunerated.

The annual staff rankings were an opportunity to coach staff in my team on this. My direct reports would come with high ranking of staff. On asking for quantifiable evidence, they would provide qualitative anecdotes. I would acknowledge but refer to the targets. I would remind them of the need to be objective and fair to all staff, pointing out that 'effort is recognized but result is what counts and to be rewarded'.

My journey from the storeroom to the boardroom was one from 'head' to 'heart'. In my early years in the storeroom, my focus – the head – was on how I could deliver superior results. However, as I grew into senior leadership roles into the boardroom, the focus became less about me, but more about others –thus the 'heart' became more important.

This supports the philosophy that success hinges on intelligence (IQ), emotional quotient (EQ), social quotient (SQ) and adversity quotient (AQ). IQ plays a strong role at lower levels of management, but EQ and SQ are invaluable at the leadership level along with AQ during challenges. Many focus on IQ as the basis of education, but a focus on EQ, SQ and AQ is key to advancement.

Leaders also struggle

Leaders have personal struggles just like everyone else. This may not be much of an issue at the lower levels (e.g. storeroom), but they become magnified as one rises to senior positions (e.g. boardroom), mainly because of the impact they may have on others (e.g. staff).

I had mine, which I continued to work on improving throughout my career. I was a workaholic, waking up at 5.00 am and staying up past 11.00 pm. An unintended consequence at the senior level was that I sometimes sent emails to staff at odd hours, including at weekends. I did not expect them to act immediately, but I later realized it put pressure on them to work odd hours themselves, as they did not want to delay. I also travelled frequently, visiting up to 20 countries in some years, and different time zones also meant odd hours for team members.

Despite coaching many staff on the need for work–life balance, I would not describe myself as a role model, which was partly why I took the decision to retire early. For many years (2008–18), I worked away from my family. I was keen for my children to have stability in schooling. I did not want to move them round the four locations where I had to work in those years. With young children, growing up fast, I was mostly an 'absentee father'. When colleagues ask what the toughest part of the job was in the boardroom or as the Global Vice President, I would say, 'Being away from family is tougher than any work challenge.'

I could not see my family after work and I had to travel just to be with them. Although we used video conferencing to keep in touch, I missed many activities (e.g. sports, meeting teachers). During Christmas, we held our 'family meeting'. I would ask whether I should continue at work or resign and be home, since the company did not have an office where we were based. My children (especially the youngest, Dara) would remind me they were in boarding school anyway and that we did spend time together during the holidays. One of their comments that always resonated with me was, 'You can be more at home, but how will we afford school fees and the mortgage?'

A leader must be conscious of the multiplication effect of their behaviours, and consciously consider the impact on staff. The effect on staff can be like a snowball, rolling down a snow-covered hillside and becoming a boulder by the time it gets to the bottom. As a leader rises to more senior positions, they must make deliberate efforts to minimize any unintended negative impact of their behaviours.

Beyond company

During the period of my journey from storeroom to boardroom, Nigeria was also going through its own journey, an important part of which was ending decades of military rule in 1999 and transiting to a democratic journey.

Nigeria's population of 190 million[1] is a huge, diverse market, by region (250 ethnic groups) and by religion. Huge oil and gas deposits, natural resources (e.g. gold), fertile land, good weather and talented people led experts in 2001 to believe it could have been a part of the next cluster of global economic giants – Brazil, Russia, India, China (BRIC). In 2014, Nigeria was included in the MINT group – Mexico, Indonesia, Nigeria and Turkey – and that year Moody's rating agency estimated that Nigeria would be among the 15 largest economies in the world by 2050, when the country's GDP is projected to exceed US$4.5 trillion from the then US$500 billion.[2]

The country has the potential for greatness. Yet, despite this huge potential, Nigeria is a contradiction, punching below its weight and with most of the citizens living in poverty due to huge infrastructure gaps (electricity, road, rail, industries) and corruption among several of its leaders.[3]

Some attribute Nigeria's lacklustre situation to its formation in 1914, which they describe as a 'flawed marriage' of regional groups that existed separately since 1800. Others point to a deep mistrust between ethnic groups that led to civil war in the 1960s. Others point to a long period of military rule, corruption, a mono-economic dependence on oil, and loss of agriculture and manufacturing capacity. As a result, there is agitation for restructuring or even a break-up.

---

[1] See https://en.wikipedia.org/wiki/Nigeria

[2] See https://businessday.ng/companies/article/moodys-nigeria-to-be-among-15-largest-economies-by-2050-with-gdp-of-4-5-trillion

[3] See www.thecable.ng/chatham-house-582bn-stolen-from-nigeria-since-independence

However, there is also progress in some areas, including digital mobile telephony and computing (85% in 2017 from 0.5% in 2001), the growth of a US$4 billion entertainment industry, online banking and reforms, NLNG becoming the fourth largest company globally and the ongoing construction of one of the biggest refineries in the world (650,000 barrels a day).

Progress on development has been slow due to historical corruption and limited revenue (low oil prices). Government has, in recent times, barely been able to meet the recurring expenditure and security spend, and has had to rely on borrowings to progress capital infrastructure development.

There are no silver bullets or short-term fixes for Nigeria's historical missteps. It needs transformation built on a long term (10–25 years) plan, executed 'brick by brick' with discipline. Reforms are required on diversification, infrastructure development, stronger institutions, education, health and the battle against corruption.

Citizens are rightfully concerned with the state of the economy and development; however, there are some citizens with a different agenda who have been spreading false narratives to undermine government. Like my experiences at NLNG, improving communication (especially on social media due to the high population of youths) and engaging citizens to correct the negative myths will be important.

In the speech I delivered at Harvard University in 2014, I posited that Nigeria needed 'dreamers of a certain kind': nation-builders, leaders with vision, integrity, courage, inclusive and able to attract talent to execute the dream. These are lessons from my journey from the storeroom to the boardroom, with relevance for the country.

Moving on

My mind drifts back to my apartment in The Hague. The balcony was a favourite spot. The Hague does not have many high-rise buildings, so the balcony provided a bird's eye view of the city. One could see far beyond the city limits. It was especially fascinating during Christmas and New Year, which were marked with fireworks at midnight. It was as if amateur pyrotechnicians wanted to outdo each other. The displays were bright and loud, went on for hours and provided a scintillating view of the city. I will miss that view!

The balcony also held other memories. It was a great place to entertain. Every summer in the years that I lived there, I hosted the Nigerian community in Hague to well-attended barbeques. Many came in bright native traditional attire and we served Nigerian delicacies, with Nigerian music playing. The event was also a mentoring opportunity and a time to interact, and to share stories and experiences.

I will miss these gatherings, but having made friends who will last a lifetime, I know that good memories never fade.

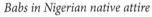

Babs in Nigerian native attire   *Babs at The Hague*
                                              (2019)

I will miss many things at Shell, especially the people. I met so many talented people at several Shell locations across the world and learnt a lot from interactions with them. I owe them so much and cannot thank them enough.

I enjoyed my 26-year career with Shell from my start in the storeroom to becoming President of a UK Institute (CIPS) and being appointed as CEO and member of the boards of NLNG and BGT, and subsequently as a Global VP, a trajectory I could never have contemplated from humble beginnings growing up on the farm. The lessons I learnt growing up helped me during the journey, including focus, determination, improvement and networks. I also stood on the shoulders of my colleagues, and learned there are rarely true impossibilities, except for those we create.

Personal integrity and courage were key differentiator values that made me stand out in the crowd throughout my career journey from storeroom to boardroom. They came across in the tough decisions I had to make, including the drive for transparency (e.g. oil spills, government revenues), community transformation (e.g. the 25-year master plan), standing up to wrongs (NIMASA), defending value (e.g. the NLNG Act), confronting corruption (e.g. handing over unscrupulous staff to security agents) and creating local capacity (e.g. the helicopter). They provide insights for global businesses operating in a developing country.

The differentiators also came with risks, including a challenge to integrity (e.g. my career threatened when I refused unethical instruction by my boss) and the need for courage (e.g. the attempt to remove me as CEO). They required personal sacrifices as relatives and friends struggled to understand why I would not award them contracts or jobs. It was tough, as the question was invariably 'Why should you be different?'

It was important not to lose direction during the journey. Core values are a lighthouse for when tough decisions are required and when no choice is clearly the better one. If one compromises on core values, then one goes nowhere!

My storeroom to boardroom journey may have ended, but I am starting a new journey now. One to fulfil my dream of spending time with my children and being a more present part of their growth. One in which I plan to continue coaching and mentoring young talents and business executives. One in which I plan to continue learning and developing myself. One in which I plan to share my experiences, in consultancy and advisory roles, with professionals in the public and private sector. I look forward to this new chapter with much anticipation!

Epilogue

God preserving me for future impact

I thank God for allowing me to be alive now and able to write this book.

When, in 1988 during national service, I was chased by marauding expelled students and attacked with a cutlass and cudgels.

I was hidden for hours before the principal and policemen arrived and rushed me to the hospital for treatment. Investigation revealed that the students had searched for me from house to house. Their intention was to finish what they started. They meant to kill me.

I am grateful to God that this brave family (Olorunmodimu) had the courage not to give me up, as without them I could have been killed. I would then not have been around in 1993 to have the opportunity to join Shell in the storeroom and enjoy a fulfilling career. I would not have worked globally and experienced that there are no impossibilities.

I would not have been around in the late 1990s to get married and have my beautiful children or to see the children grow, with the first now a graduate of Imperial College London and the others following behind.

I am so glad that God preserved me, as if He had not, I would not have been around in the 1990s, when the conscience of Nigerians was aroused following the military annulment of an election. I would not have seen the people rally to oust the military in 1999. I would not have been around to see civilian rule from 1999, including a transition from one party to another.

I would not have been here in 2000 to transfer to the United Kingdom for seven years. I would then not have worked with other Christians there to build a parish of the Redeemed Church

Fountain of Love, or to see the great Christian movement growing in Scotland.

Without the act of kindness of the family, I would not have been here in 2008 to become a Director of SPDC and later Non-Executive Director of West Africa Gas Pipeline and Vice President of Shell Sub-Sahara Africa. I would then not have been part of developing local capacity in Nigeria, which has grown in leaps and bounds.

If not for the family's courage, I would not have been here in 2011 and been appointed by the Board of Nigeria LNG as MD/CEO of NLNG, the largest plant in sub-Saharan Africa. I would also not have been here in 2016 to become a member of the Shell Global Upstream Leadership Team in Netherlands, and a Vice President across 40 countries

I would not have been able, in 2014, to lead an organization that straightened its back up and refused to pay illegal levies despite being blocked. I would not have been here to hear the court rule in 2018 that we were correct. I would not have been part of arousing the conscience of the nation and demonstrating what ethical leadership and courage are about.

I would also not have been able to go to Harvard University and tell Nigerians of the need for dreamers of a certain kind: inclusive nation-builders with integrity, able to rally the talents in the country towards a vision, and not distracted by pernicious myths.

I would not have risen to become a Global President of a UK Institute. I would not have gone to speak at conferences across the world and tell professional colleagues of a dream that I had, of the profession raising its game and raising its voice.

But for God's kindness, I would not have been here in 2015 to see the Bonny community rally around a 25–year Master Plan. I would not have led NLNG to embark on impactful projects including the university laboratories, Bonny–Bodo Road, NLNG Head Office and to pursue Train 7 growth.

I am glad that God preserved me, as if He had not, I would not have been here in 2020, when the Nigeria youths rallied, using social media, to demand an end to police brutality. I would also not have witnessed the government accepting in record time the yearnings of citizens. I would also not have been around to see the nation's leadership conscience aroused.

I thank God for that family hiding me in their cellar. Without their act of courage and integrity, I would not have risen from the storeroom to boardroom, and this book would not have been written.

I thank God for His mercy.

Printed in the USA
CPSIA information can be obtained
at www.ICGtesting.com
JSHW012030140824
68134JS00033B/2971